T0227584

A practical guide to Single Storey House Extensions

by Andrew Williams

Routledge
Taylor & Francis Group

LONDON AND NEW YORK

For

Geraldine . . . my wife

and

Thomas Grenville Williams . . . my late father

First Published 1989
By Taylor & Francis

Published 2004 by Routledge
2 Park Square, Milton Park, Abingdon, Oxon OX14 4RN
711 Third Avenue, New York, NY 10017, USA

First issued in hardback 2017

Routledge is an imprint of the Taylor & Francis Group, an informa business

ISBN 13: 978-1-138-40894-4 (hbk)
ISBN 13: 978-1-85032-033-3 (pbk)

Publisher's Note
The publisher has gone to great lengths to ensure
the quality of this reprint but points out that some
imperfections in the original may be apparent

contents

from the author

Dear Reader,

This book was first written as an 'in-house' manual for use by my assistants. I have now published it for use by other Surveyors/Building Consultants/Draughtsmen and people wishing to design plans for their own extensions. As you will note I have also included appendices which cover such things as suggested conditions of contract, suggested specifications, and standard letters for those people who wish to take their first step towards creating their own part-time business. These documents have been written by myself over the years to cover various eventualities. I have found that they have served me well but I am not suggesting that they are foolproof and would only hold them out to be examples and suggestions. I cannot enter into correspondence or accept liability for any loss or consequential loss from those persons who choose to use the information that I have provided. Where standard terms and conditions and/or specifications are included these should be modified as necessary to suit the type of project as obviously all the clauses may not be applicable to every situation.

For those people wishing to use the information as a basis for a part time business I would suggest that they note in particular the first 19 or so clauses in Appendix 'F'. These clauses provide protection to the building consultant and client. I would suggest that something along these lines be included on all your plans or as a standard specification document. REMEMBER in the majority of cases you will not be supervising the work on site and unscrupulous builders will try to blame you for their own inadequacy IF YOU LET THEM.

I have called this book 'A Practical Guide to Single Storey House Extensions' because in my experience the vast majority of small extensions are of a single storey nature and I believe that to burden a beginner with the problems of staircase design and the like would be too much in the first instance. It is hoped that it will be possible to follow this book with more advanced information at some future date.

I have presumed that the reader will have the basic knowledge of technical drawing/ draughtsmanship and has used a drawing board at school or technical college. If not, then it is essential that you take further advice from a friend or colleague who has a knowledge of draughtsmanship prior to purchasing expensive equipment.

I hope that after reading my book that you will be able to confidently tackle simple single storey plans and to be able to submit them for Planning and Building Regulation Approval and deal with queries raised by Building Control and Planning. Obviously, a little knowledge can be a dangerous thing and I would not envisage that my readers would rush out and try to design multi-storey extensions and buildings of major importance. It would also be advisable for you to practice your new skills by submitting plans for extensions to your own house, a friend's house or a close relative, before acting for paying clients. I must stress that it would not be possible to condense all the information available on the matter of Building Control and Planning into one volume. (Even if I knew all the answers.) The purpose of this book is to enable the average lay person to confidently tackle simple Planning and Building Control procedures in connection with small works without falling into the most common traps but the reader must be prepared to make further enquiries into problems which might not be envisaged in this book. IF IN DOUBT CHECK AND DOUBLE CHECK.

A. R. WILLIAMS

introduction

Making a start

So you want to design and construct an extension on a property. Where do you start? What are the procedures?

For small extensions there are two main controlling bodies, the local planning department and the local building control officer. These departments are completely separate even though in many cases they come under the Head of Technical Services at your local council. More information on these two departments is given later.

The first thing that one must appreciate is that even though you own your own house or your client owns the house there is very little that one can do to it without consulting your Local Authority and your Building Society. The best way that I can explain the procedures is to give an example of a badly handled case.

An example of an incorrect approach to planning and building control

I shall call the lady concerned Mrs 'X'. Mrs 'X' rang me at my office and asked me to visit her to discuss the problems that she was having with the Local Authority. Listening to her on the phone, one would have thought that the Council officials concerned were relations of Count Dracula. From what Mrs 'X' told me she had carried out alteration work to her house without bothering to obtain Building Control or Planning Approval. (These terms will be explained later.) She could see no reason why Planning and Building Control were necessary but could I help her.

After discussions with the Local Authority, I informed Mrs 'X' that I would advise a retrospective Planning and Building Control Application. However, I also let her know that I would not guarantee that this would be accepted by the Authority. N.B.: In my experience it is never prudent to promise what you can't guarantee. Although you prepare plans for your client you are a human being and not superman (or woman) and although some people advertise and say 'Planning and Building Control approval guaranteed', I find it difficult to know how those individuals can make such claims.

So what was this alteration that Mrs 'X' had had carried out? She had removed the up and over garage door to her integral garage and had installed a Georgian bow window with brickwork below and then turned the garage into a storeroom with a connecting door. Why were the Local Authority objecting to the alteration?

Well firstly, because the alteration work materially affected the external appearance of the property the council believed that they ought to have been consulted (that is plans should have been submitted before the work was carried out). Secondly, the council had adopted policies relating to domestic properties in this area which required that each dwelling should have a garage or parking space and the council

4

did not wish to 'open the floodgates' by allowing people to convert garages, on this housing estate, into living accommodation and then find that cars were left parked in the street. Also, as the garage only had half brick walls (not cavity walls – see later for details) the alterations did not comply with current building regulations. Mrs 'X' had a problem. Where did she go wrong?

The correct approach to local planning and building control

Mrs 'X' could have visited her local planning office/building control department or consulted someone like myself before the work was put in hand. I have always found that the Planning and Building Control Officers are very helpful if approached in a reasonable manner and will always discuss your proposals and help with planning, listed building consents, tree preservation orders etc and discuss Building Control requirements. I would stress that these people are very busy. With government cutbacks in some authorities they are very understaffed but if you make an appointment you will receive good treatment. If you turn up unannounced they could all be out on site. One of the most important things to appreciate is that various local authorities have differing local policies to suit the problems in their areas. These policy decisions are usually published in a 'House Extension Policy' – (See Appendix A) or some similar document. Note: Appendix A is only given as an example and you must get a copy issued by your Local Authority if they have one.

What is the difference between planning and building control?

This is a question I get asked a great deal by clients. The easily understood answer is the Planning Department are interested in what the building looks like and Building Control want to know if it is structurally sound. To put it another way, planning is not an exact science. The Planning Department can accept or reject an application on 'feelings' – it didn't like the scheme for some reason. (It is a little more scientific than that – usually based upon the problems of the area.) Building Control deals with facts. For example, an extension must be provided with a damp proof course at 150mm (6″) above ground level or the 'U' value of an external wall of a habitable room must be minimum 'U' value of 0.60 W/m² °K. These details will be explained later.

More details on building control

Building Control in the British Isles is not standard. The comments that I am making are restricted to 'The Building Regulations 1985' which apply to the whole of England and Wales except London. London, Scotland and Ulster have their own building control system. For people living in the Isle of Man they also have their own regulations. The power to make building regulations in England and Wales is vested in the Secretary of State for the Environment by Section 1 of the Building Act 1984 and these change from time to time so you must ensure that you remain up to date. The Building Regulations have recently undergone a major change (11th November 1985). Under regulations 11 and 12 it is possible to adopt two procedures which are as follows:

1. Send a building notice to the local authority.

2. Deposit full plans.

Owing to the problems that could reasonably be anticipated with the first option – Building Notice – I would not recommend this procedure except in very exceptional circumstances. You as the designer could be found liable in negligence if work carried out on the site does not comply with Building Regulations. The idea of the system is however that it will speed up building operations in as much that work can start on site virtually immediately.

This book however assumes that you will adopt the second option – Deposit Full plans – as this system protects both yourself and your client against contravention of the Building Regulations. I have however included a question and answer document (see Appendix G) which answers some of the points on the Building Notice approach to Building Control.

The new regulations are much shorter than the old regulations (24 pages) but now refer back to approved documents (approximately 300 pages) and has a manual with sketches.

Certain extensions are excluded from Building Control requirements and are listed in Schedule 3. The ones most likely to be encountered by my readers are as follows:

1. Greenhouses;

2. Temporary buildings or mobile homes;

3. Class VII buildings – not exceeding 30m². i.e. Greenhouses, conservatory, porch, covered yard, covered way, carport open on two sides.

N.B.: Just because the Building Regulations don't apply to these structures under the new regulations **does not mean that planning permission** is not required. I would advise the reader to send a copy of their plan to Planning and Building Control and ask for confirmation that Planning and Building Control consent is not applicable – be safe not sorry. If you require more details than this book contains I would suggest that you visit your library and borrow 'The Building Regulations Explained and Illustrated' – published by Collins. It is an excellent book but it will take some reading as it contains over 500 pages of information.

Why is building control required?

The only way that I can explain the need for some form of building control is by way of anecdote. The story that I am about to relate is perfectly true and unchanged in any way.

I was asked to visit a property by a small building contractor and duly did so. My brief was to design a small porch on the front of the house in question. Normally, I restrict comment to the matter in hand (i.e. the porch) but when the lady of the house let me in to measure up I immediately enquired about the chimney breast which had obviously just been removed from the living room.

I said, 'Have you put in R.S.J's (Rolled Steel Joists) to support the remaining chimney breast upstairs?'

The husband looked at me and shook his head. I then looked around the rest of the house and realised that he had demolished all the internal downstairs walls and that the entire weight of the first floor was supported on a 4″ × 4″ (100mm × 100mm) wooden pole and the external walls. I warned them that I considered the whole property structurally unsound and left very quickly. People who have no knowledge of building construction should not dabble.

The gentleman in question made it clear that he could see no reason for submitting plans to the council for the work he had carried out.Neither could he see anything wrong in his 'Do-it-yourself' attempt. However, when the house collapses, as it will, and the next door neighbour sues him, I dare say he will.

It is people like him who will save money at any cost and thereby create a danger to themselves, their families and neighbours, that makes Building Control so essential.

What approvals do I have to obtain when building an extension?

Subject to the above exemptions, what approvals does the average householder have to obtain before building an extension? I will be discussing the following matters in greater detail later but generally speaking when you are building an extension to a house the application to the Local Authority will be in one of two forms:

Typical application type 1

1. A letter to the Planning Department (enclosing a copy plan), requesting confirmation that the proposals are **permitted development.**

2. A Building Control Application.

Typical application type 2

1. A full Planning Application.

2. A Building Control Application.

The details of how you go about making a Building Control and/or Planning Application are the subject of following chapters.

Advice on comprehending the facts in this book

It is doubtful if you will be able to absorb all the facts on simple Planning and Building Control procedures from one reading of this book but I would advise you to read it right through once before commencing drawing your plans.

I have presented the Building Construction chapters first because most people want to understand the construction details before they even start to consider the submission of plans for approval.

However, it is essential that you realise that certain procedures exist before commencement of your plans so that you ensure compliance with all the rules.

building construction

Foundations

The foundations of the house are probably the most important part of the property (every part is important) because without an adequate base the property will quickly become unstable.

There are several types of foundation for domestic construction which are:

1. Traditional strip

2. Deep strip

3. Tied footing or raft

4. Piled (very unlikely to be used)

Traditional strip

A typical traditional strip footing is indicated on Fig 1. Study the sketch carefully because you will have to use similar details on your drawings. The width of the concrete base must comply with the Table E1 of Approved Document 'A' (page 22). As it is unusual for a client to allow trial holes to be dug (pits about 3′ 0″ square and 3′ 0″ deep dug in the garden to test soil conditions), I normally indicate on my plans that trial holes have not been taken and that the Builder is to test as necessary. A minimum width is then indicated as 600mm on the plans as this is a convenient size for a bricklayer to work off and it allows for most ground conditions.

It is essential that the ground underneath a foundation is stable enough to carry the imposed loads. If for any reason you suspect or know that the ground on which the proposed extension is to be built is **'made up'** (i.e. **fill or hardcore**) or that the area is an old tip or that trees have been growing on the land which will have disturbed the subsoil, then you must consult your Engineer for a **Raft Design (or other non-standard footing** see later for details) because a 'Building Control' standard foundation assumes that the ground, upon which the 'Traditional Strip' footing is built, will be undisturbed. Walls must be built centrally on a concrete footing (i.e. about the same centre-line). The building inspectors will reject your plans if walls are built 'off centre' unless you can prove that the non-traditional foundation is acceptable.

Note also that it is sometimes necessary for the builder to use sulphate resisting concrete. Sulphates in the ground (whether natural or man made) attack the concrete and it will eventually disintegrate. As sulphate resisting concrete is more expensive than concrete made with ordinary Portland cement, I am not suggesting that you specify it in all circumstances but I am recommending that you put on your plan 'Sulphate resisting concrete if necessary'.

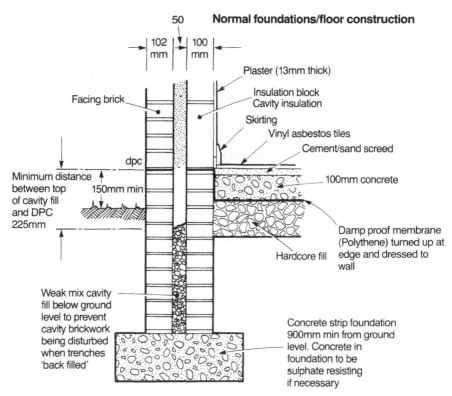

50

102 mm

100 mm

Plaster (13mm thick)

Insulation block
Cavity insulation

Facing brick

Skirting

Vinyl asbestos tiles

Cement/sand screed

dpc

100mm concrete

Minimum distance
between top
of cavity fill
and DPC
225mm

150mm min

Damp proof membrane
(Polythene) turned up at
edge and dressed to
wall

Hardcore fill

Weak mix cavity
fill below ground
level to prevent
cavity brickwork
being disturbed
when trenches
'back filled'

Concrete strip foundation
900mm min from ground
level. Concrete in
foundation to be
sulphate resisting
if necessary

Fig 1 Typical domestic foundation/concrete floor

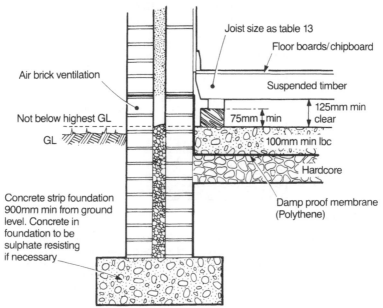

Joist size as table 13

Floor boards/chipboard

Air brick ventilation

Suspended timber

125mm min
clear

Not below highest GL

75mm min

GL

100mm min lbc

Hardcore

Concrete strip foundation
900mm min from ground
level. Concrete in
foundation to be
sulphate resisting
if necessary

Damp proof membrane
(Polythene)

Fig 2 Suspended timber floor

This brings to the builder's attention that he must check the ground conditions before building. (It also covers the designer against claims from a possible irate client.)

The concrete must be of a minimum consistency (minimum 50kg cement to not more than 0.1m³ of fine aggregate and 0.2m³ of coarse aggregate or grade C15P to B.S. 5328:1981 or grade 15 concrete to CP110:1972) and the minimum thickness of a foundation must be 150mm (6″) and that the thickness of a foundation must not be less than the projection. So if the concrete projects 200mm for some reason, the thickness must be 200mm. Where there are changes in level (stepped footing) the 'stepped' section must have a minimum thickness of 300mm. If piers are indicated on walls then the footing must be enlarged to give the same projection from its face as the footing is from the main wall. **Note also** the details indicated on Figs 4,5 and 18 which are self explanatory.

Deep strip

Similar to traditional, but deeper and probably 'trench fill' as Fig 3 – Used near trees. **Note** also Fig 18. It is recommended that you refer to NHBC practice note 3 if there are trees nearer than 7.5m to the proposed extension foundation.

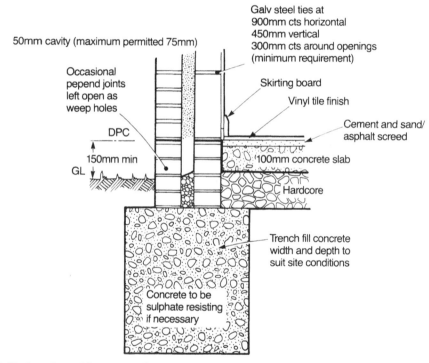

Fig 3 Modern deep strip or trench fill foundation

Some provision for water to weep from the base of cavities is advisable.

The width of foundations can be checked with the building regulations but under normal conditions a width of 600mm is practical if a bricklayer is to stand on the foundation to lay his bricks/blocks.

N.B. Details are not to scale, for clarity the general notes have been spread around Figs 1, 2 and 3 eg tie wires on Fig 3 apply throughout, so does note regarding 225mm gap between dpc and top of cavity fill.

Foundations near drains

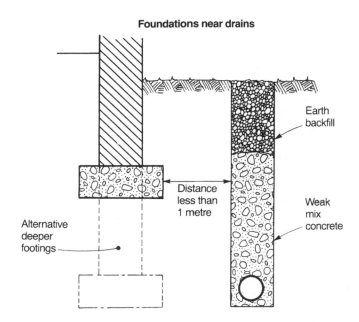

Earth backfill

Alternative deeper footings

Distance less than 1 metre

Weak mix concrete

Fig 4

Where a drain is closer than 1m of a foundation and the trench is lower than the wall foundation then the trench must be concrete filled up to the underside of foundation.

Where the foundation is over 1m away and the drain is deeper than the foundation then the concrete fill shall be distance-150mm as indicated in Fig 7.

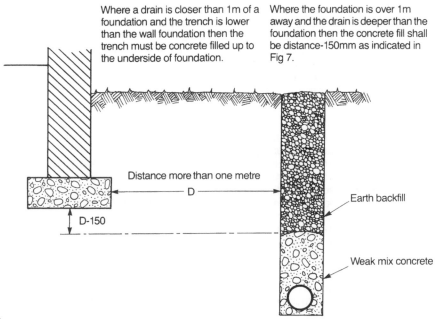

Distance more than one metre

D

D-150

Earth backfill

Weak mix concrete

Fig 5

The tied footing or raft

A tied footing or raft foundation is a very useful form of foundation but does not come within the scope of the Building Regulations. The simple form of tied footing or raft which is met within domestic construction usually comprises of a reinforced floor slab with a toe beam (see edge detail on raft – Appendix B). Because it is outside normal Building Control parameters, calculations are required by most Building Control departments before raft foundations can be used. A set of simple calculations are provided in Appendix B at the back of the book which should cover most domestic situations and treats the edge beam as a tied foundation (i.e. it assumes that the steel in the slab and toe prevent overturning movement caused by non-central loading), But, if you are in doubt consult a structural engineer.

The object of the true raft foundation is to spread the load of the structure above onto a very wide area of ground. Rafts are usually required where ground conditions are known to be bad or variable. The raft or tied footing indicated in Appendix B can also be a useful form of construction where it is necessary to build up to a boundary line but the next door neighbour will not allow the spread of the footing to pass under their garden.Unless you are able to provide the council with calculations the assistance of an Engineer is required. You will note from my standard conditions that I exclude the cost of Engineer's fees from my fee. Experience has taught me that this is a wise precaution because whilst it is relatively easy to assess ones own work load the likely Engineering charge is unknown. Should the reader decide to offer design services after reading my book and wishes to obtain the services of an Engineer, try advertising in the local paper.

Walls

The walls below ground level usually comprise two half brick skins with a weak concrete cavity fill. Note that the cavity fill is kept down 225mm from the wall damp proof course. External walls to habitable areas must attain a 'U' value of 0.60 W/M²K. I usually specify a facing brick outer skin, 'Dritherm' cavity insulation and 100mm Thermalite or Celcon aerated concrete block inner skin of the cavity wall. Under the 1976 Building Regulations, window areas were required to be at least 10% of floor area of each room but not more than 12% of the wall area unless double glazed. The ventilation areas were also supposed to be 5% (1/20th) of the floor area. The new 1985 Regulations have dispensed with the 10% rule but I still indicate this on my plans because I consider it a sensible window to floor ratio.

Lintels

The most usual type of lintels used on modern extensions is metal 'Catnic' or similar type. Try to obtain a manufacturers' catalogue and study the details. These lintels have been designed by the manufacturer to take normal loads above door and window openings. If you are in any doubt about the strength of any lintels specified, most manufacturers will check the loadings for you free of charge if you send them a copy of your plan. (**Check that there is no charge prior to sending the plan.**) For cavity walls it is usual to use the 'Combined' type which provides a cavity gutter over the opening as well as providing support. **Always** provide adequate bearing at the ends of lintels. The manufacturer's recommendations should be followed but the usual bearing is 150mm (6") each end. **If you don't give enough end bearing the lintel will fail under load.**

Should you wish to use concrete lintels a schedule of suitable types is given in the 'Builders Reference Book', together with details of steel reinforcing bars required for various spans.

Rolled steel joists (RSJ's)

Where one is forming an opening in an existing wall to connect the new extension to the existing it is normal practice to install two RSJ's bolted together and seated on adequate concrete padstones. Usually two 7″ × 4″ RSJ's will do for most openings in an 11″ cavity wall up to 2 metres wide, in two storey properties, but it is unwise to assume that this is the case. Once again the Building Control department of most local authorities require Engineer's Calculations.

Describing structural calculations is beyond the scope of this book but the calculations are not difficult. You therefore have three alternatives which are as follows:

1. Ask your structural engineer to provide you with several typical sets of calculations for typical openings.

2. Obtain one of several books from your local library and work them out yourself.

3. Use a computer.

N.B.: A good program is produced by W.L. Computer Services which runs on the BBC B/Master (Tel: 051 426 9660) but some reading around the subject is advisable if you have no knowledge of the theory of how a steel beam works.

Roof

A typical flat roof is detailed on Figs 6 and 6a. Don't forget the vapour barrier!

The size of roof joists can be determined from schedules contained within approved document A pages 8 to 33.

The roof detail indicated is a 'Cold Roof' which is the cheapest type of flat roof. The roof is covered usually with three layers of bituminous felt (built-up felt) which is covered by a 13mm layer of white limestone chippings laid shoulder to shoulder in

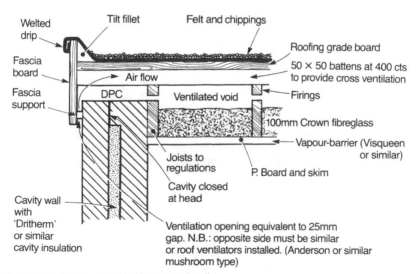

Fig 6 Ventilated cold roof with Abutment detail

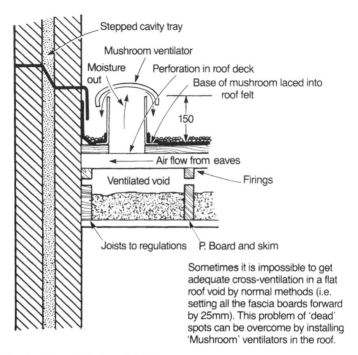

Stepped cavity tray

Mushroom ventilator

Moisture out

Perforation in roof deck

Base of mushroom laced into roof felt

150

Air flow from eaves

Ventilated void

Firings

Joists to regulations   P. Board and skim

Sometimes it is impossible to get adequate cross-ventilation in a flat roof void by normal methods (i.e. setting all the fascia boards forward by 25mm). This problem of 'dead' spots can be overcome by installing 'Mushroom' ventilators in the roof.

Fig 6a Ventilated cold roof (Abutment detail)

bitumen. The purpose of the chippings is to reflect the heat of the sun. Without them the roof will perish. The other alternative is to use solar reflective paint but as this is not accepted by most Building Control departments as having a fire rating, solar reflective paint finish can't be used near to boundaries with other properties.

N.B.: Flat roofs are not flat, they have a fall of not less than 1:40. Also note that a minimum of three layers of felt must be used. Where a prefelted chipboard is used the prefelting must **not** be counted as one of the layers. The only reason that prefelted chipboard became popular with builders was because 'blacktop' once on, and the joints taped, it provided protection from the weather and other trades could carry on working inside the building (e.g. plasterers) while they waited for the felt roofer to come to the site. With the cold roof, it is essential that the voids above the insulation, be vented to the external air. Eaves ventilation should be provided by a gap of 25mm left between the fascia board and top of wall. **Condensation** will form on the roof deck and will cause staining on the ceiling below and eventually rot the roof joists and roof deck if adequate ventilation is not provided. Besides, the new 1985 Building regulations have specifically covered condensation because it is now a major problem.

N.B.: If the client is agreeable to additional cost, a plywood base to the roof is far superior to a chipboard one.

Another form of flat roof widely used is the warm roof but I don't tend to use this form of construction because of cost. Some clients quite rightly are worried about the watertightness of Flat Roofs – at a little extra cost the problem can be overcome by using H.T. felts as opposed to BS 747 felts. If you ring Anderson's Roofing they will be able to supply details of their High Tensile felts and details of specifications for various situations.

Pitched roof

Refer to detail Fig 7. A pitched roof comprises a covering of tiles and/or slates. These must be laid to a suitable pitch (i.e. not less than the manufacturer's recommended pitch). If you are using a pitched roof, ring Marley, Redland or other roof tile manufacturer and obtain their catalogue. This will give you a complete specification. The roof tiles and slates are fixed to battens over sarking felt. The whole structure is supported by rafters, the size of which can be found in the schedules in the Building Regulations. For roofs of any reasonable span, the ceiling joists act as ties and prevent the roof spreading. The size of ceiling joists can also be obtained from the Building Regulations. Fig 7 only shows a roof of small span but on larger spans it is necessary to insert purlins and once again details of sizes can be obtained from the Regulations. As with flat roofs, 100mm of glass fibre must be incorporated between the ceiling joists and laid over the plasterboard soffit (ceiling).

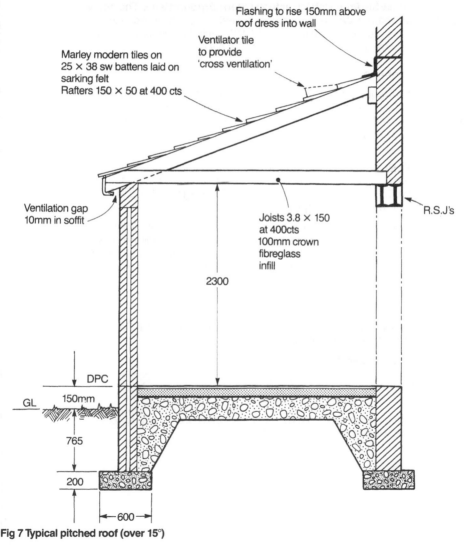

Flashing to rise 150mm above roof dress into wall

Ventilator tile to provide 'cross ventilation'

Marley modern tiles on 25 × 38 sw battens laid on sarking felt
Rafters 150 × 50 at 400 cts

Ventilation gap 10mm in soffit

Joists 3.8 × 150 at 400cts 100mm crown fibreglass infill

R.S.J's

2300

DPC

GL   150mm

765

200

600

Fig 7 Typical pitched roof (over 15°)

Solid floor

The most common type of floor construction being used in England and Wales is the solid floor. This comprises of a hardcore bed (no thickness is specified but a minimum of 150mm (6″) is practical). For those who do not know, hardcore is crushed brick or crushed stone levelled and compacted as a bed for concrete.

It is essential that rubbish is not used as hardcore. The hardcore must not contain sulphates, bits of wood, plaster or other impurities which could deteriorate or attack the concrete bed which is laid on it. The hardcore is **blinded** with fine material, usually sand, which prevents the 1200 gauge polythene from being punctured (see Figs 1, 2 and 3). A 100mm 1:2:4 mix concrete bed is then laid on the polythene and a 50mm cement and sand **screed** laid on top of that. The alternative to a cement and sand screed is flooring grade asphalt and a typical specification is incorporated in the appendix of this book. The vinyl or vinyl asbestos tile or carpet is then laid on the screed. **Do not show the floor slab lower than external ground level. The house will flood in wet weather.** Where the hardcore beneath a solid floor exceeds 600mm in thickness you are advised to refer to NHBC practice note 6, and comply with the requirements otherwise the concrete floor slab will settle.

Suspended timber floor

The timber floor still requires a hardcore bed and 100mm concrete slab (see Fig 2) which is why it is more expensive to construct in normal conditions. (If a large amount of fill is required because of site levels it could be cheaper to put in a timber floor.) Some people prefer timber floors as they are not as hard on the legs if one is standing up a great deal.

The **honeycomb sleeper walls** (i.e. walls built so that they have holes left in them to allow airflow) are constructed off the concrete slab and wall plates and joists built off them. Note the D.P.C's and essential dimensions. **Once again – keep the concrete subfloor surface above ground level.**

N.B.: The voids underneath timber floors have to be provided with ventilation to the outside air. This is done by installing air bricks, but they should not be built in too low, otherwise water will flood into the house in wet weather. The purpose of air bricks beneath a timber floor is to prevent dry rot spores germinating on the wood of the floors.

Note on dry rot

Dry rot is a fungus that literally eats wood and it has a terrific growth rate. To germinate, true dry rot needs three things:

1. Moisture in the wood over 20%.

2. Bad ventilation.

3. Warmth.

Air bricks prevent the conditions arising where dry rot can flourish. If you specify tanilised timber this will also help to stop woodrot outbreaks.

matching materials

The Objective

When an extension is completed, it should look like part of the original and not like an extension.

One of the problems that small extensions create is matching new materials up with existing.

In the main, as the designer of very small extensions, this possibly will not be your problem but if you are asked to supervise the work to completion, it certainly will be. Here are a few tips.

Matching Bricks

There are several problems with matching bricks to the existing. The first is that colours fade and become stained over the years and even if it is possible to find the exact brick match, the result may be unsatisfactory. The second is that even bricks produced by the same manufacturer vary in colour. The final problem is that many brick manufacturers cease production either because they work out their supply of clay or because they go out of business. In order to get over this problem, if it is possible to get a good sample of an existing brick, then take it to a 'Brick Library'. You laugh . . . but there are such things.

Whilst they are not common, large builders merchants and specialist brick suppliers do have such things and what is more are usually prepared to let you have sample bricks to show your client.

The next problem with bricks is size. In days gone by, bricks came in a variety of sizes (3in, 2in etc). The modern brick is usually 65mm thick (2⅝in) plus or minus a few millimetres. What chance have you got of matching the new work to the old if the old walls are in unmanufactured 3in bricks. The answer is none, unless you can find a brick that gives a reasonable match in your brick library or can obtain 'second-hand bricks'.

Second-Hand Bricks

There are some suppliers that specialise in obtaining, cleaning and reselling 'second-hand bricks', but I only know of one in my area. However, it would be worth making enquiries. Correct use of second-hand bricks can overcome a large number of problems.

Matching mortar

It is possible to obtain pre-mixed coloured mortar. Several firms manufacture it. (Tilcon in my area is probably the most common.) These companies will usually supply a small pack of mortar if asked to do so. The mortar comes in a wide variety of shades and can usually be made to imitate the mortar of the existing house. The 'rough stuff' that is supplied however has to be mixed with cement before use. Ask the supplier to advise on the amount of cement required.

Bond

There are several types of bonding for brickwork.

I have indicated the most common types on the adjacent sketches (Figs 8, 9, 10 and 11). In cavity walls it is usual to nowadays use stretcher bond.

But what if the existing house wall is solid construction and built of English Bond or Flemish Bond?

The way around this problem is to use 'snap headers'.

In other words your external cavity wall will still only be half brick thick (4.5in) but the bricklayer will chop the headers in half to imitate the old brickwork that you need to match.

Fig 8 Stretcher bond
The external appearance shown as all 'stretchers'

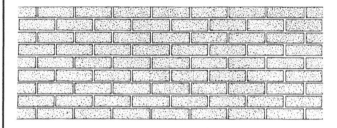

Fig 9 English bond
The external appearance shows alternate layers of 'headers' and 'stretchers'

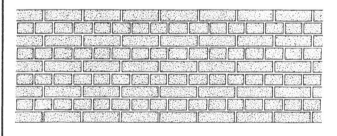

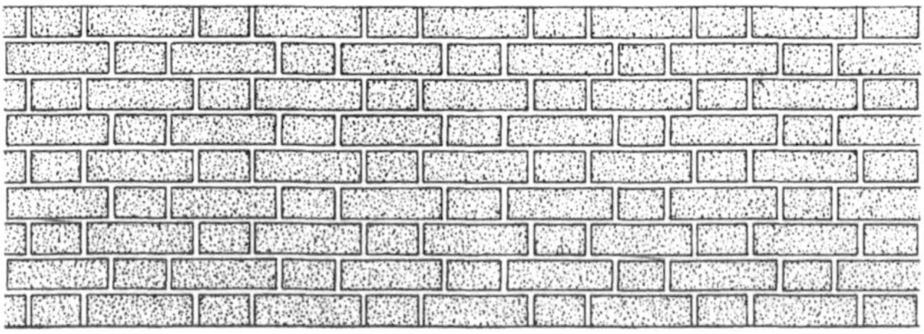

Fig 10 Flemish bond
The external appearance shows alternate 'headers' and 'stretchers' in each course

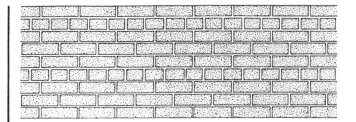

Fig 11 English garden wall bond
The external appearance shows mainly 'stretchers' with a row of 'headers' every three, five or seven courses

Vertical abutments

Where new walls abut old walls the brickwork should be cut, toothed and bonded to the existing. In other words, pockets cut in the old brickwork and the new bricks let in.

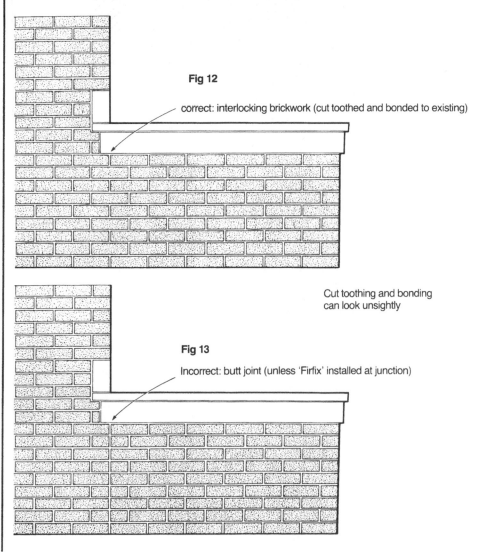

Fig 12

correct: interlocking brickwork (cut toothed and bonded to existing)

Cut toothing and bonding can look unsightly

Fig 13

Incorrect: butt joint (unless 'Firfix' installed at junction)

Note also sketch 14, which shows a way to avoid cut, toothing and bonding showing at corners.

The only way of avoiding cutting and toothing is by using patent jointing strips such as Furfix which is bolted to the existing wall and has steel 'teeth' that bed into the mortar joints of the new brickwork. The manufacturers instructions should be followed when using any product such as this, especially the instructions regarding mastic pointing. (To keep out water.)

Fig 14 Set back at corner

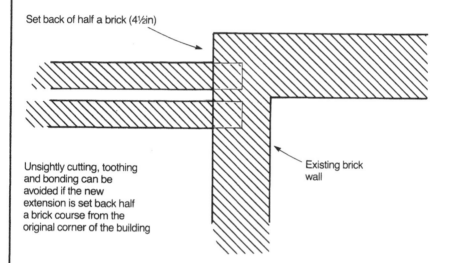

Set back of half a brick (4½in)

Unsightly cutting, toothing and bonding can be avoided if the new extension is set back half a brick course from the original corner of the building

Existing brick wall

Matching tiles

Tiles are usually more difficult to match than bricks but we are only dealing with single storey extensions here. As it is not usually possible to see both the extension roof and the main roof at the same time unless you stand a long way away from the house to be extended, it is usually possible to match the tiles in colour only and the finished work not to look unsightly.

building regulations

Generally

As previously indicated the Building Regulations were revised in 1985. These regulations are to control HOW a building is constructed and NOT what it looks like.

When the regulations were revised the objective was to SHORTEN THEM and make them simpler.

The regulations now are shorter because there are less of them but what has happened is that they now refer to subsidiary documents and taken as a whole the building regulations are far longer and more comprehensive than any previous document.

The list of building control documents is as follows (if obtained as a package from H M Stationery Office), High Holborn, London WC1V 6HR (or their approved suppliers):

(1) Manual to the Building Regulations 1985.

(2) Approved Documents to support Regulation 7.

(3) Approved Documents A.

(4) Approved Documents B.

(5) Approved Documents C.

(6) Approved Documents D.

(7) Approved Documents E.

(8) Approved Documents F.

(9) Approved Documents G.

(10) Approved Documents H.

(11) Approved Documents J.

(12) Approved Documents K.

(13) Approved Documents L.

(14) Mandatory rules for means of escape in case of fire.

I am not going to attempt to deal with the content in detail because it would be beyond the scope of this book but I have attempted to provide an outline to give my readers some idea of what these documents contain.

(1) The manual

This document outlines which buildings are covered by the building regulations. As I have indicated before some buildings are outside the building control system. (See the introduction chapter for details.)

It also covers the two systems of building control (ie, full plans or the building notice. See introduction chapter for details).

The manual then states the main regulations and refers the reader to the various Approved Documents. It is a useful document because alongside each regulation. (Printed in green) there are explanatory notes which help the reader to understand the true meaning.

(2) Approved document to support regulation 7

This document covers materials and workmanship generally, short lived materials, resistance to moisture, use of High Alumina Cement (HAC).

(NB – Please do not specify HAC anywhere on your drawings except as a heat resisting material – the use of HAC has been the cause of collapses at swimming baths etc, and has caused local authorities a great deal of expenditure – its use was banned in France at the turn of this century but in Britain we seem slow to learn).

The document also covers the problem of House Long Horn Beetle and the districts most affected.

(3) Approved document A

This document amplifies regulation A1 and A2. It says basically that the building must be designed so that it won't collapse or become unsafe. It then goes on to provide definitions, types of timber, sizes of floor joists, ceiling joists, binders, purlins, rafters, etc, and the approved spans. It also covers thicknesses of external walls and the use of straps to ensure lateral support.

This document is full of sketches which indicate how work should be constructed.

I would advise the reader to obtain and study approved document A as this will be a massive help with plan production and understanding basic construction.

(4) Approved document B

This covers fire spread in buildings and is mainly concerned with larger buildings. Note however, the requirements of garages (see page 10 of approved document B). The wall and any floor between a garage and house, where it is attached to the house, must have a 100mm step where there is a door opening. This opening must have a half-hour fire door with a door closer fitted.

Also note pages 42 and 43 of the approved document B. This gives details of notional fire resistance periods for variation type of construction.

Approved document C

This covers site preparation and resistance to moisture. It basically amplifies regulations C1, C2 and C3 and covers the removal of top soil under buildings, dangerous and offensive substances and subsoil drainage. I would recommend that the reader obtains and studies the document as it discusses siting the damp proof courses, floor construction, ventilation under suspended timber floors, installation of cavity trays, and cladded external walls. There are a large number of diagrams which will be helpful.

Approved document D

This deals entirely with cavity insulation and specifies the relevant British Standards for cavity wall insulation. The main concern is that no cavity insulation should give off toxic fumes.

If you specify Fibreglass 'Dritherm' as I do then you should have little trouble with the local authority as fibreglass is inert.

Approved document E

This covers airborne and impact sound and amplifies regulation E1, E2 and E3.

When designing small extensions this document will not be referred to a great deal.

Approved document F

This document covers ventilation and condensation and is an essential document to study.

Condensation has become a major problem in new housing. The document covers window ventilation (natural ventilation) and ventilation to pitched and flat roofs (see pages 3 and 6 of approved document F). I have covered these basic requirements elsewhere in this book.

Approved document G

This document covers hygiene generally (ie, food storage, bathrooms, hotwater storage and toilets). The most important rule that affects small extensions is that WC's (toilets) should not open directly into a space used for food preparation (including a kitchen).

The way around a toilet area opening into a kitchen is to install an 'air lock' or ventilated lobby. This comprises of an area approximately 1 metre $\times$ 1 metre with two doors between the WC and the kitchen.

Approved document H

This covers, drainage generally (ie, soil and vent pipes, cesspools, rainwater, and foul drains).

I would advise my reader to study pages 3, 4, 5, 8 and 12 of approved document H in particular as the sketches are virtually self explanatory.

Approved document J

This covers heat producing appliances. Note in particular Table 1 of approved document J. This requires that a permanent air entry opening with a total free area of at least 5500mm² or 50% of the appliance throat opening has to be provided to the outside air.

This obviously only applies where an old appliance is affected by the alteration or where a new one is being installed. Page 4 covers chimney heights and flue sizes. Page 6 gives dimensions of hearth sizes etc.

Approved document K

This covers staircases etc. As this book only deals with single storey extensions it is outside the scope covered, but the document is worth studying by those wishing to tackle more advance work.

Approved document L

This is a vital document and covers heat loss in buildings. As I have indicated previously roofs must have a U value of 0.35W/m2K and walls 0.6W/m2K in dwellings.

This document describes in detail how to ensure that the extension complies IF THE LOCAL AUTHORITY ASK FOR HEAT LOSS CALCULATIONS. THEY RARELY DO if the construction complies with their recommendations.

Mandatory rules of escape in case of fire

In most cases this will be beyond the scope of this book.

Regulation changes update

At the time of writing there are consultative documents in circulation from Eurisol the UK mineral wool trade association recommending a change in U values.

It is proposed to reduce the U values of walls from 0.6 to 0.45W/m2K and domestic roofs from 0.35 to 0.25W/m2K.

The reader is advised to keep himself updated on these.

single stack system

General Notes

As anyone knows, who has studied physics at school or has ever made home-made wine, if you want to get liquid out of one vessel into another it is possible to make a simple syphon with a length of tube. Once the syphonage has started the liquid will empty into the second vessel without any problem as the weight of the water descending in the tube will suck the liquid out of the first vessel.

As you will probably be aware, most sanitary fittings and sinks have a 'Trap' underneath which prevents foul smells entering your house. The principle of syphonage caused a problem for plumbers because sometimes on a long run of pipe, syphonage would start and empty every trap in the house thus defeating them.

This problem was overcome originally by putting in additional ventilation pipes, the sole purpose of these being, to let air into the system and prevent syphonage occurring.

Plumbing

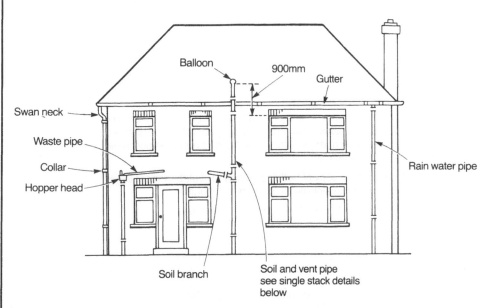

Balloon

900mm

Gutter

Swan neck

Waste pipe

Collar

Hopper head

Rain water pipe

Soil branch

Soil and vent pipe
see single stack details
below

Fig 15 Names of external plumbing fittings etc

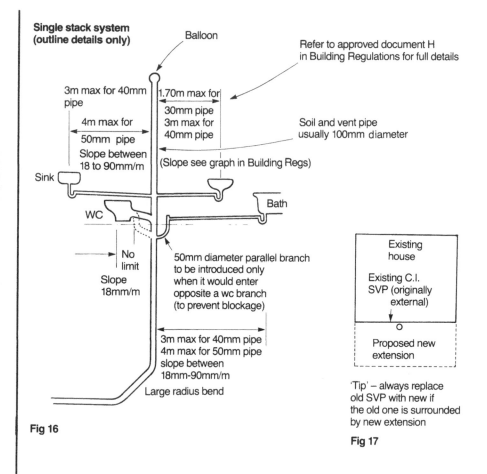

Single stack system (outline details only)

Balloon

Refer to approved document H in Building Regulations for full details

3m max for 40mm pipe

4m max for 50mm pipe

Slope between 18 to 90mm/m

Sink

1.70m max for 30mm pipe
3m max for 40mm pipe

(Slope see graph in Building Regs)

Soil and vent pipe usually 100mm diameter

WC

No limit

Slope 18mm/m

50mm diameter parallel branch to be introduced only when it would enter opposite a wc branch (to prevent blockage)

Bath

3m max for 40mm pipe
4m max for 50mm pipe
slope between 18mm-90mm/m

Large radius bend

Fig 16

Existing house

Existing C.I. SVP (originally external)

Proposed new extension

'Tip' – always replace old SVP with new if the old one is surrounded by new extension

Fig 17

If one looks at the back of one or two of the older buildings in your town, it is still possible to see examples of 'Two pipe' systems (i.e. the ones with secondary anti-syphonage pipes above the larger soil pipes).

The problem of providing anti-syphonage pipes everywhere has been overcome nowadays by using the 'single stack' system of plumbing on soil and ventilation pipes on dwellings **but** the system works to very specific limits and rules and these have to be observed or the system will fail.

It is suggested that you obtain a copy pf Building Research Establishment Digest 249 and approved document H as this gives full recommendations.

If you look at Fig 16, I have given a simplistic version of the details.

Practical 'Tip'

Sometimes a new extension will be built so that the old soil and vent pipe which was originally outside the building now becomes an internal pipe. Don't just show the pipe boxed in. Indicate a new plastic patent system to be installed (e.g. Key Terrain, Hunter Plastic etc) because the old cast iron type SVP could leak and cause damage at some future date (see Fig 17).

kitchen planning

Generally

A well planned kitchen undoubtedly increases the value of the property. I know, because I sold my house almost entirely upon the kitchen. To be honest the rest of the house didn't have a great deal to recommend it – except for the decoration.

However, unless I am requested to do so by my client, I do not provide internal kitchen layouts on my plans. Most clients prefer to deal with this themselves but from time to time it is necessary to carry out this service.

A kitchen is a workplace and like most workplaces people in the past have examined what makes the best layout. The best layouts obviously make life easy for the person working in the kitchen (i.e. keeps walking distance to the minimum while allowing adequate room to move around), provides adequate storage space and enough work tops.

Kitchen planners use what is known as a work triangle to make a kitchen as ergonomically efficient as possible. This triangle links the three main areas of the kitchen, the fridge/fridge freezer, the sink and the cooker.

Ideally, when you add up the sides of the work triangle the distance should not be less than 3.60m (11' 0") and not more than 6.60m (21' 6"). If the dimensions are any bigger than these, it will result in unnecessary walking, if they are any smaller the kitchen will be cramped and awkward to work in.

Do's and Dont's (Trying to do your best)

1. Try and work out a kitchen layout on a 600 × 600 grid. (Most units are designed to a module.)

2. If possible put in a double bowl sink and place it under a window.

3. If possible keep the sink near the soil stack. It is recommended that the distance be 2.30m maximum.

4. Don't put the sink in a stupid position (i.e. in a corner of a room or against the side of a tall cupboard). If the sink has to be close to an abutment try to put a short length of worktop in 100-300mm long to allow some 'elbow room'.

5. Don't site the sink so that the 'dryer up' can't get in or where the cook risks scalding the 'dryer up' if something has to be taken off the stove quickly. Put the bowl side nearest the cooker and not the drainer side.

6. Don't put cookers in front of windows or under wall units. There is a risk of fire! Allow space around the cooker for the door opening – allow at least 300mm from an abutment so that the door can be opened fully.

7. Don't site a sink, cooker etc, so that someone entering the room will bang the door into the person working in the kitchen.

8. Don't put the cooker and sink on opposite sides of the room. Don't put the cooker and sink in a position where a door comes between them, otherwise if the door is opened quickly someone could get burnt.

9. The washing machine is best near the sink – it cuts down on plumbing.

10. If there is a tumble dryer don't have it away from an external wall or the house will get full of steam. Tumble dryers should be permanently vented to the outside air if possible.

11. Try to allow space for a fridge and an upright freezer and allow opening space on the hinged side (100mm min).

12. If possible put an extractor fan in the kitchen.

13. Work Tops – try to link the appliances with runs of work tops.

Electric

Most kitchens have too few power points and lights. It is recommended that the kitchen has what is known as background lighting and also task lighting. However, I do not intend to cover this aspect as it is the province of kitchen planners proper, but when it comes to power points, sensible provision can be made.

14. A cooker (electric) should have its own 30 amp (minimum) supply from the distribution board. It should not be taken off the ring main. The control should be at the side of the cooker and not over it.

15. Electric sockets in kitchens should be mounted at least 150mm above worktops. **Note** if they are too low there is a risk of water causing an electric fault.

16. Where possible always specify double socket outlets in a kitchen and put in an adequate number, one for each appliance (and in a modern kitchen there are a large number).

17. Where you have to install switched spurs, put a neon light on them so that everyone knows when the power is on.

For further information

Refer to NHBC recommendations and BS 3705, Department of Environment Design Bulletin 24 and Parker Morris Report 1961.

trees
and other obstructions

Tree and other obstructions

When designing an extension to a house a site visit is always essential. In my experience there are two major items to be careful about which are as follows:

The unknown drain/electric cable

So you visit a house on a large housing estate and there is an unaccountable gap between your prospective client's property and the next house or he has an unusually large garden. Beware! Builders never leave large bits of ground out of the goodness of their hearts. They will try to get as many houses on their plot of land as possible.

Your client then tells you that he has this drain in his garden. **Check!** (see Fig 18).

Is it a drain or is it a sewer? The water board own sewers and quite often will place restrictions on the ground above them. In my area, you are not allowed to build within 3m from the edge of the sewer.

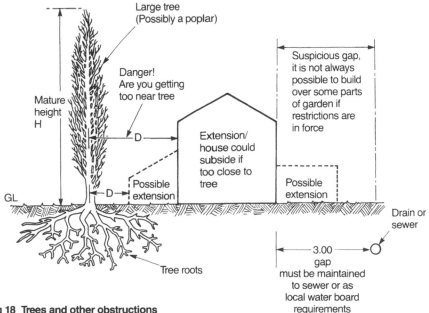

Fig 18 **Trees and other obstructions**

If you visit your Council Offices they can usually tell you if there is a sewer running across a piece of land. Large electric cables installed by the local electricity board can also cause similar problems.

If the obstruction just turns out to be a deep drain then the only problem is ensuring that the foundations of the house will not subside into the drain. This can be resolved by concrete encasing the drain or by putting in deeper foundations to the extension (see Figs 4 and 5).

Trees

The NHBC classify trees into two groups, Poplars, Elms, Willows and any other type.

The first NHBC group are harmful when seen in relation to houses. All trees close to houses can cause damage but Poplars are particularly bad because their roots run everywhere, sucking moisture out of the ground and on clay soil can cause subsidence to normal strip footings. Near trees, deep footings must be constructed to Building Control approval.

The NHBC recommend that the following formula be adhered to when building a house or extension near trees, and traditional footings are used, and the house or extension is nearer to the tree than two thirds of its **mature** height or **full size.** (See Fig 18.)

(D = Distance of centre of tree from house. H = Expected mature height.)

Type of Tree	Ratio of D/H						
	1/10	1/4	1/3	1/2	2/3	3/4	1
	Depth of foundation trench in metres						
Poplars, Elm or Willow	N/A	2.80	2.60	2.30	2.10	1.90	1.50
Any other tree	N/A	2.40	2.10	1.50	1.50	1.20	1.00

drawing the plan

Generally

I would recommend that you study my standard plan.

The plan is not meant as a solution to all situations but to provide good grounding. If you produce a drawing something like this you are well on the way to getting approval to your scheme.

Note well

The Building Control Department very rarely approve a plan without raising a query. To be fair, they are protecting society against itself. How many people do you know who would try to 'get away with something' if they could. Considering how much money is spent on the average extension, it never ceases to amaze me how mean and petty some people can be over small costs. The earlier anecdote is an example of this.

Basic considerations

The details listed below are requirements of the Building Regulations.

Scale
The plans shall not be drawn at a scale of less than 1:100 except the block plan which shall not be less than 1:1250 and the key plan not less than 1:2500. It is essential that the scale that you are using is shown on all plans and that a North Point is indicated. It is not normal on small extensions to have to provide a block plan and key plan, in my experience. (You must provide one or the other.)

N.B.: Take a compass with you when you visit the property to be extended and note North.

Details
You must provide a plan indicating 'as-existing' for all floors being altered and elevations and on some occasions it will be necessary to indicate plans of all floors whether altered or not. A plan 'as-proposed', elevations 'as proposed' and sections through the existing and extension/alteration are also required.

Basic faults of beginners – 'Old Chestnuts'

1. Cavity walls

The modern form of construction for external walls of habitable rooms is a cavity wall. That is a wall built with two skins tied together with wall ties. I normally specify that the cavity walls shall be composed of an outer brickleaf, 50mm cavity and on the inside, a leaf of 100mm Celcon or Thermalite Blocks. The cavity is normally 50mm (2") wide and can be filled with glass fibre batts or polystyrene to add thermal insulation. N.B.: Cavity Insulation is purpose made and normal fibreglass cannot be used. If you have a copy of my plan read carefully my specification for cavity walls. I specify 'Dritherm' which is impregnated with chemicals to waterproof it.

2. Cavity closings

Note that your description must include closing cavities at 'heads' and at door and window openings. Closings at doors and windows must incorporate vertical and horizontal damp proof courses (DPC). See Figs 19 and 20 for details.

3. Wall ties

There should be a minimum of 5 No. wall ties per m². I tend to specify vertical twist ties as opposed to 'butterfly' ties because they are stronger and likely to last longer.

4. Garage walls/porch walls

A half-brick (102mm thick) wall can be used in garages and porches but piers size 225mm × 102mm must be provided at maximum 3.00m centres in garages as strengthening.

5. Flat roofs

There are two main types of flat roofs used on small extensions – the warm type and the cold type. I have described them before but I will repeat the basics. I mainly use the 'cold type' of roof because it is generally cheaper to construct. Note my description for the flat roof. Don't forget to provide 13mm of limestone chippings bedded shoulder to shoulder in bitumen or solar reflective paint. (This stops the sun damaging the roofing felt.) One must be careful when specifying chipboard. There is a roofing grade and if standard chipboard is used it will disintegrate after 12 months (if not sooner). If your client can afford it specify WBP bonded plywood for the roof deck.

6. Roof insulation and ventilation

The roof must contain 100mm of glass fibre insulation or another approved insulant. Voids above the glass fibre must be ventilated to the outside air so that moisture is 'sucked off'. Note the airway between the fascia board and wall.

7. Roofs with pitches less that 15 degrees

They should be cross ventilated at eaves level on opposite sides of the roof and the vent area should be equal to a continuous gap of 25mm on each side. This can be achieved by setting the fascia boards forward by 25mm. Where 'dead spots' are formed in roof voids and 'cross ventilation' cannot be attained, one can install Anderson flat roof ventilators (see Fig. 6a). These are like hollow plastic mushrooms which sit on the roof – obtain a catalogue from Andersons Roofing.

8. Ventilation

The new 1985 regulations no longer require the ceiling heights to be 2.30m and it is also permissible to ventilate one room from another as long as there is a permanent opening of not less than 1/20 of the combined floor area.

9. Soakers and flashings at abutments

These are specially made pieces of lead (or approved substitute) which fill the gap between the roof tiles and an abutting wall. See Figs 20 and 22.

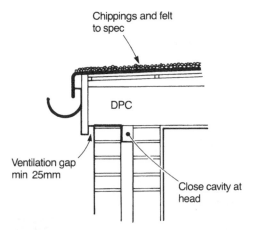

Chippings and felt
to spec

DPC

Ventilation gap
min 25mm

Close cavity at
head

**Fig 19 Cold roof details with
ventilated fascia**

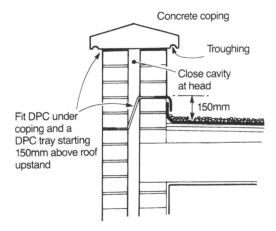

Concrete coping

Troughing

Close cavity
at head

150mm

Fit DPC under
coping and a
DPC tray starting
150mm above roof
upstand

Fig 20 Parapet wall detail

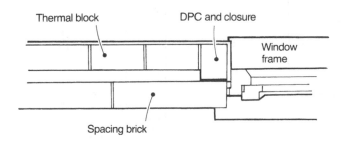

Thermal block

DPC and closure

Window
frame

Spacing brick

Fig 21 Closure and DPC on vertical

Sundry details

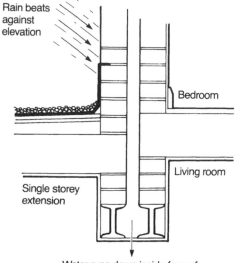

Rain beats against elevation

Bedroom

Living room

Single storey extension

Water runs down inside face of outer leaf and penetrates between rolled steel joists

Fig 22 Incorrect detail for steel beams

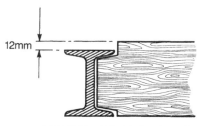

12mm

Joists trimmed into steelwork must project at least 12mm above the top of the steel section to allow for timber shrinkage. Ensure that the bottom faces are level

Fig 23 Trimming to steelwork

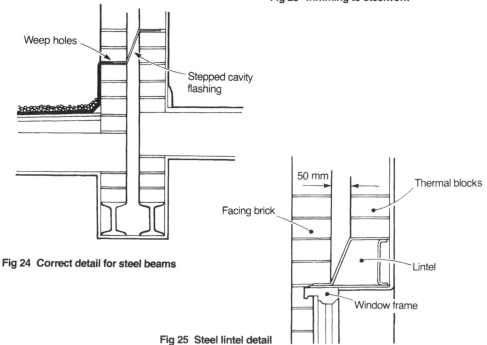

Weep holes

Stepped cavity flashing

Fig 24 Correct detail for steel beams

50 mm

Facing brick

Thermal blocks

Lintel

Window frame

Fig 25 Steel lintel detail

10. Ground floor

There are two forms of ground floor construction in general use. The timber suspended floor and the concrete solid floor. The concrete solid floor is the cheapest and the most widely used. This is why it is shown on my plan. If you have my plan, note the description on the standard specification.

Don't forget the polythene 1200 gauge DPM even if you use asphalt screeds. It is also essential that the visqueen is 'turned up' at abutments and joined to the DPC in the wall. If you don't do this the extension will suffer from damp penetration.

11. Ground levels

The damp proof course in an external wall has to be a minimum of 150mm above ground level externally. It is normal practice to make the floor level at the same height. **Never** show a floor level below ground level. The house will get flooded in wet weather if it is actually built that way.

12. Foundations

See previous chapter for full details.

13. Drainage

This can be a problem on any extension because sometimes the drains are not always in the positions that you think they are. Try to locate the drains as existing as best you can and lift any manhole covers. In order that the drains can function the falls or slope of the bottom of the pipe must be adequate but not too great. Why? If it is too great the water will run away and leave the solids behind to block up the drains. If the slope is too shallow the drain will not empty at all. It is a rule of 'Thumb' that (100mm) 4" drains are laid to falls of 1:40 and (150mm) 6" drains are laid to falls of 1:60. **All drains under building should be encased in concrete 150mm thick.** Put this note on your drawings if you don't the inspector will query it. Under the 1985 Regulation all drains must have flexible joints. (E.G.: Hepseal or Hepsleve pipes.)

14. Structural stability around windows/openings

For those who have obtained a copy of 'The Building Regulations' I would suggest that they refer to them. The regulations require that the number of windows in a wall will not impair its structural stability (or for that matter any other wall).

Windows must have brickwork on each side equal to one sixth of the width of the window. Where two windows are close together then the wall has to be one sixth of the two windows lengths added together. No opening should exceed 3.00m. If it does engineers calculations will be required.

What a simple plan should have on it

For a typical 3.00m × 3.00m rear extension the following minimum details would be required by planning/building control are:

(a) Plan of the existing ground floor – minimum scale 1:100;

(b) Plan of proposed ground floor (existing house and extension);

(c) Section through building – minimum scale 1:100;

(d) Existing rear elevation or front elevation (as applicable) – minimum scale 1:100;

(e) Proposed rear elevation – minimum scale 1:100;

(f) Site location plan – minimum scale 1:1250;   Sometimes these

(g) Block plan – minimum scale 1:2500.   are combined.

It is recommended that scales of 1:50 to be used wherever possible as this makes the plans clearer. Obviously, if the proposed extension is more ambitious then a larger number of plans, elevations and section will be required. If the extension or alterations affect two floors then plans for both floors must be provided (proposed and existing).

Check list of items which drawings should have on them

Below are a list of common faults to watch for:

1. Under the 1976 Regulations a minimum floor to ceiling height of 2.30 metres was required. Under the 1985 Regulations there is now no floor to ceiling restriction (except for headroom on staircases) but I always indicate a minimum headroom of 2.30m where possible.

2. The floor level/oversite concrete level should not be lower than outside ground level.

3. The damp proof course (DPC) in the walls should be minimum 150mm above ground level.

4. The floor levels in the extension should indicate that they are at the same level as the existing.

5. Indicate opening vents on windows.

6. Windows over 12% of wall area should be double glazed.

7. Windows should be minimum 10% of floor area. (Now no longer compulsory.)

8. Opening lights on windows should be minimum 5% of floor area.

9. When building across a driveway with a garage and there is no other access around the house always provide rear door.

10. Windows in existing house which open into a proposed attached garage should be blocked up to prevent a fire hazard.

11. Where a 'room' opens off a 'room' and there is no other means of light/ventilation the minimum opening between the two must be 1/20 of combined floor area.

12. As 'rule of thumb' drains should fall to a gradient of 1:40 for 100mm drains and 1:60 for 150mm drains. (However note new formula in approved document H.)

13. Where drains and sewers are close to proposed new foundations ensure that foundations are low enough not to impose any load on the drain or sewer. Note also Figs 4 and 5. If a deep drain is near to a wall the trench must be backfilled with concrete. Note formulas.

14. Returns to windows should be at least one-sixth of the window opening it abuts. Where between windows one-sixth of combined window.

15. Provide night vent of 10,000mm² to all patio doors. Between garages and houses provide half-hour fire doors complete with suitable frame having 25mm rebates, door closer (e.g. two No. Perko or one No. Briton). Provide 100mm step to prevent petrol coming into dwelling in the case of fire.

16. Normally two No. 7″ × 4″ R.S.J's bolted together will provide adequate support to new openings in external one brick (9″) or cavity walls (see Figs 24) up to spans of 2.00m. However, it is best to ask your Engineer to check the strength

by calculation if you are in any doubt. Some Local Authorities always ask for calculations anyway.

17. Remember that an R.S.J. is only as strong as the wall supporting it. Provide adequate padstones at each end to spread the load. (Allow 225mm × 225mm × 150mm minimum.) N.B.: Thermal blockwork can have a stress value as low as 0.28 N/mm². Design for the worst case.

18. Always be on your guard for local peculiarities and interpretations of the Local Building Control Department or Planning Department. If in doubt ask you Local Authority.

19. When specifying windows, doors, baths, kitchen fittings etc, obtain catalogues (keep them in a filing system for frequent use). Particularly useful catalogues are Magnet and Southerns/Boulton and Pauls.

Equipment required by the surveyor when carrying out a survey

The following equipment is recommended whilst carrying out a survey. It has been compiled after many years of carrying out 'on-site' surveys:

1. Stout A4 clip board;

2. Adequate supply of paper (note some people prefer using graph paper for this purpose);

3. Compass for checking north point;

4. Variety of coloured Biros. (This makes it easier to distinguish dimensions from the outline of the Building);

5. Instant camera. (This saves a great deal of time for both the draughtsman and for the surveyor when 'on site');

6. Two metre folding surveyor's 'Staff'. (I find that a double sided metric is best);

7. Manhole keys. (Heavy and light duty recommended);

8. Screwdriver and jemmy for Manholes;

9. Crow bar.

Modern methods of producing drawings – the word processor

When I was in my teens I often assisted my father with the preparation of small plans and I very quickly found the process of writing and rewriting a standard specification onto each and every drawing extremely time consuming and boring. Now, this is one aspect of plan production which does not worry me. A small computer with an adequate keyboard and a printer can easily help to save the tedium of reproducing a standard specification. There are many small computers on the market but to give you some idea of the 'set up' which I have. I will briefly describe how I word process my drawings.

I have a BBC Model B Computer with a 'wordwise' processor installed and this is linked to a TEC 25 printer. The advantages are simple. I have a standard specification on disc (you can easily use cassette). The details are quickly transferred into memory and any amendments quickly made. I reckon that I save an hour and a half on each and every drawing by using my word processing system. The specification also looks better and there is little chance of missing off an important clause.

Note: For those interested, a standard specification pre-recorded on tape or disc can be provided. For prices please write or phone W.L. Computer Services (051 426 9660).

planning and building control

When is planning consent required?

Recent Town and Country Planning Acts gives Local Authorities the power to control 'Development'. In other words most items of work on a property are controlled in one way or another unless they happen to fall within the categories referred to as 'Permitted Development'.

The important exceptions which affect small works are Class 1 and Class 2 permitted development. Planning Permission for an extension to a dwelling house is **not** required under Class 1 permitted development if **all** the following criteria are met with **(not applicable in National Parks, Areas of Outstanding Beauty and Conservation Areas and Listed Buildings – see latter):**

1. The height of the existing building is not increased by the extension.

2. The extension is not built at the front of the house facing the road. (Note porches under 2m² floor area.) The cubic content (measured on external walls and roof) of the extension is not more than 50m³ on terraced properties or 70m³ on other properties. There is a supplemental rule for larger properties which allows a 10% addition to terraced or 15% to other properties (percentages apply to floor areas) up to a maximum of 115m³.

3. No part of the building as enlarged, improved or altered which is within 2m from a boundary is to exceed 4m in height.

4. More than 50% of the garden must be left after you have built the extension.

5. The extension does not obstruct a view of the highway.

6. A porch as long as the floor area does not exceed 2m² is not more than 3m high and is not less than 2m from a highway (back of pavement) and as long as it does not cause loss of visibility to motorists.

N.B. If a private garage is built in the back garden and there is a 5m gap between the front of the garage and back of the house then that garage cubic capacity will not be added to the cubic capacity of another extension (as long as the 5m gap is maintained).

The operative date for calculating what the original dwelling house was, is 1st July 1948. In relation to buildings constructed after this date it means the date of construction.

Obviously some people might if following the above rules decide to extend their house by 70m³ this year and another 70m³ the year after. **You cannot do this.** The planning acts work on the principle that all measurements start in 1948.

Once you have used up your permitted limit – that's it.

How does permitted development assist an application?

If you are preparing plans for a property and you believe that the extension proposed is permitted the advantages are as follows:

1. The client will not have to formally apply for Planning Permission which speeds up the construction process.

2. The fees charged (currently £33.00) will not have to be paid.

However beware

I never take the risk of informing a client that the extension is automatically exempt.

Let the local authority tell **you** that it is!

How to deal with Class 1 permitted development

Because there are so many provisos (see in particular my notes on Conservation Areas and the like) I usually advise a prospective client when he is likely to be allowed 'Permitted Development' Status for an extension and then write to the Local Authority along this line:

Bigtown Planning Department,
Municipal Offices,
Somewhere Road,
Hyville.

For the attention of Mr. Slowbend

Dear Sirs,

Proposed extension at 22 Bridge Street, Somewhere

Please find enclosed Plan No. .... This indicates a Flat Roof extension, not exceeding 70m³ with a height of 3 metres. There have been no extensions to the property (as far as we are aware).

We are of the opinion that this extension constitutes Class 1 Permitted Development and would request your confirmation/comments in writing.

Yours faithfully,
J. Soap.

Doing this has three effects:

1. It covers yourself and your client. If you advise that an extension is permitted and it is not – then you are in trouble.

    Let the Local Authority decide.

    It is vital to remember that Local Authorities can grant conditional planning permission (as they have done in my area on large densely developed housing estates). As part of the permission they can prevent future permitted development (P.D.). Also see my previous comment on Conervation Areas etc.

2. It saves the client the planning fee. Under recent legislation Local Authorities charge for Planning and Building Control Applications.

3. It saves about two months in time if the authority confirm that P.D. Status is acceptable.

The other important exemption is given under Class 2. This allows minor operations such as the erection of gates, fences, walls or other small enclosures as long as:

1. The height above ground level does not exceed 1m if it abuts a highway.

2. There is no obstruction of vision on a highway.

Listed buildings – (Town and country planning act) (listed buildings and buildings in conservation areas) Regulations 1977

If a property is listed it means that demolition, alteration and additions will be allowed only after proposals have been carefully examined and alterations or extensions must not deface the character of the original. If a building is listed by the Department of the Environment then there is an obligation on the Local Authority to notify the owner and occupier as soon as their house is placed on the list. It is unlikely therefore that a client will be unaware that his house is listed.

The demolition or alteration of a listed building without prior permission is punishable by imprisonment and/or a fine of £1,000.00.

When dealing with an alteration to a listed building a separate application in addition to planning application must be made to the Local Authority. I would suggest that you contact your Local Authority and ask for an application form for alteration to a listed building so that you can study the contents. Alterations to listed buildings have to be advertised in Local Newspapers giving details of the proposals and saying where the plans may be inspected. The public have 21 days in which to make observations to the local planning authority. A notice giving similar details must be posted on the site.

Conservation areas

The Civic Amenities Act 1967 gave statutory recognition to Conservation Areas and a duty to Local Authorities to determine which part of their area is of special architectural and historical interest. There are many different kinds of conservation area (e.g. a whole village, the centre of a town, a terrace, a small group of buildings).

Local Authorities are required to create and publish schemes of enhancement for conservation areas. Grants may be available to individuals who live in conservation areas – it is worth checking with the Local Authority. Expenditure on Professional fees also qualify for grant aid. Grants are however not available for new extensions.

N.B. In conservation areas permitted development rules do not apply.

When dealing with work in a conservation area you must be prepared to make enquiries with the Local Authority regarding their policies.

Tree preservation orders

A Local Authority may make a tree preservation order relating to any tree, group of trees or a belt of woodland and these may be in fields, gardens or building sites. Hedges are not covered but large trees growing in hedges could be.

This is one of the reasons why the planning application form asks if any trees are to be felled.

Note on planning and building regulations

Until you obtain sufficient experience it is advisable to speak (in person) to both the Planning and Building Control Department before you start work on your drawings. They will however want you to bring sketches with you so that you can make yourself understood.

It is also useful if you take a small piece of tracing paper with you when you visit the Local Authority offices. The reason that I say this is because some Authorities will let you trace a site location plan from their ordnance survey sheets. (Any drawing submitted to the Local Authority should have a location plan with a North point so that they can locate your property.)

Be prepared however for a nasty shock . . . Ordnance Survey decided that they were losing money by allowing their plans to be traced and now some Local Authorities are charging several pounds per A4 copy of an ordnance sheet. You can either pay the money or do as I sometimes do . . . measure the garden up and produce your own approximate sketch of the locality. Take a compass with you for your North point.

meeting the client

Meeting the client

From experience the prospective client wishing to have plans drawn up telephones either in response to your advertisement in a local paper or by recommendation. It is therefore unlikely that you will know your client and I consider it essential to take with me a small form of agreement (see appendix C). Why you might ask? It only takes a few bad payers to realise why this is essential. My guiding rules are:

1. Agree scope of work.

2. Tell the client what you do and what you don't do.

3. Agree the price.

4. Get him to sign on the dotted line that he wishes you to proceed (see appendix D).

It is also useful to carry a standard survey sheet so that you don't forget important matters:

Standard survey sheet

1. Name of Client ....

2. Address of Client ....

3. Owner Y/N

4. If tenant address of owner ....

5. Does the alteration affect neighbour by having footings projecting into the next door garden? Is a letter needed?

   NB
   Examine Figs 1, 2 and 3 and you will note that the concrete footings are wider than the wall. If you build the extension on the boundary the footings will be under the neighbour's garden. This is acceptable but only if the neighbour agrees (preferably in writing).

6. Name and address of neighbour ....

7. **Instructions**

   A ....

   B ....

   C ....

   D ....

   E ....

   F ....

   G ....

8. Where are drains/manholes?

9. Has the rear garden got over 50% of area left after building extension? Y/N

10. Has the house been extended before? Y/N. If yes explain to client about the existing extension was built after 1948 that the cubic capacity counts towards 70/50m³.

11. Are there any trees close to the proposed extension? Y/N

submitting plans

Filling in the forms

Any Building Control or Planning Application has to be made on the relevant forms and obtained from your Local Authority. Each Authority's forms vary slightly. In order that you can follow the details given below I would suggest that you ring the Planning Department in your area and ask for:

1. A set of Building Control Application forms.

2. A set of Planning Application forms.

3. A copy of certificate A & B.

4. Copies of their explanatory notes.

5. A list of their fee charges for Planning and/or Building Control.

The forms may seem daunting to fill in at first and if you are in any doubt, take your forms into the Local Authority personally (together with the copy plans and relevant fees). Some Authorities have 'Householder Planning Application' forms which are easier to fill in – ask if they do!

The first few questions on both the planning and building control forms ask obvious questions such as your client's name and address, the agent's name and address and these should be simple to fill in on both sets of forms. If you are acting as Agent for someone else you must fill in your name and address otherwise all correspondence will go to your client and you will not have the correspondence to hand to deal with queries raised.

The Planning and Building Control forms then ask you the address applicable to the application. This address is not necessarily the same as the client's address. **Note the planning forms ask for the site plan to be edged in red.** The planning forms usually ask for the site area. For domestic applications I usually say **see plans.**

The planning forms will then ask for details of the development, whether the owner has interests in other land in the area, and details regarding access to the highway. See below:

1. **Applicant** (in block capitals)	**Agent** (if any) to whom correspondence should be sent (in block capitals)
Name *C.U. STOMER* Address *'DUNROMIN' CHURCH STREET* *SOMEWHERE* Tel. No.	Name *J. SOAP* Address *114 BIGTOWN ROAD* *SOMEWHERE* Tel. No.

2. **Particulars of proposal for which permission or approval is sought**

(a) Full address or location of the land to which this application relates. SITE AREA (Edged red on plan) *'DUNROMIN' CHURCH STREET, SOMEWHERE*

(b) Brief particulars of proposed development including the purpose(s) for which the land and/or buildings are to be used. *GARAGE AND PORCH EXTENSION TO EXISTING DWELLING*

(c) State whether applicant owns or controls any adjoining land and if so, give its location. (Edged blue on plan) *NO*

(d)	Does the proposal involve:-		State Yes or No	vehicular	pedestrian
		Construction of a new access to a highway.		*NO*	*NO*
		Alteration of existing access to a highway.		*NO*	*NO*

The building control forms ask slightly different questions:

1. Description of work (this is similar to planning). Purpose for which the building extension will be used (Domestic Construction is Purpose Group 1 – Small Residential) and if existing.

2. Means of water supply (Mains supply, well etc.), and mode of drainage. The planning forms also ask similar questions regarding drainage later on in the forms.

A typical extract from a building control form is given below:

1. Address or location of proposed work	'DUNROMIN' CHURCH STREET SOMEWHERE
2. Description of proposed work	GARAGE AND PORCH EXTENSION TO DWELLING
3. (a) Purpose for which the building/extension will be used	PURPOSE GROUP 1 — SMALL RESIDENTIAL
(b) If existing building state present use	AS ABOVE
4. Means of water supply	MAINS SUPPLY
5. Mode of drainage (a) Foul Water (b) Surface Water	} MAIN DRAINAGE SEE PLANS

The next question on the Planning forms regards the type of Planning Application. For simple extensions 99% of the time you will apply for **full planning permission**. See below for typical example:

3. Particulars of Application (see note 3. "Notes for Applicants")   If "outline" tick which of the following are reserved for subsequent approval.

(a)   State whether this application is for:

		Yes or No		
(i) Outline planning permission		**NO**	1. Siting and/or layout. ☐	3. Means of access. ☐
			2. Design and external appearance. ☐	4. Landscaping. ☐
(ii) Full Planning permission		**YES**		
(iii) Approval of reserved matters following the grant of outline permission.		**NO**	If yes, state the date and number of outline permission. Date: Number:	
(iv) Renewal of a temporary permission or permission for retention of building or continuance of use without complying with a condition subject to which planning permission has been granted.		**NO**	If yes, state the date and number of previous permission and identify the particular condition(s). Date: Number: The condition(s)	

The question on a Full Planning Application form will ask regarding current use of the building or land. The form was designed for use with very large projects. I usually fill this section in as follows:

4. Particulars of Present and Previous Use of Buildings or Land

State:

(i)   Present use of buildings/land: (i)   DWELLING

(ii)   If vacant, the last previous use and date when last used, if known: (ii)   N/A

for additional information which is usually filled in thus:

5. Additional Information

		State Yes or No		
(a)	Is the application for Industrial, office, warehousing, storage or shopping purposes?	**NO**	If Yes, complete Part 2 of this Form (see Notes for Applicants)	(N.B. certain small developments may not require this information please consult the Planning Division).
(b)	Does the proposed development involve the felling of any trees?	**NO**	If yes, submit plans indicating the correct position and approximate size of all existing trees on the site, identifying those it is intended to fell.	

(c)   (i) How will surface water be disposed of?   } MAIN DRAINS — SEE PLANS
      (ii) How will foul sewage be dealt with?

(d)   State the maximum height of the proposed building above ground level........ APPROX 3m........

Obviously if any of the answers given above are incorrect in your particular application then give the correct answer. If you are chopping trees down – say so – those trees might have a preservation order on them.

You will then be asked to list the plans:

$$BST/1$$

7. I/WE hereby apply for planning permission/approval for the proposals described in this application and the accompanying plans.

Date ...... **FEB 1989** ...... Signed ...... **BUILDING SERVICES (TECHNICAL)**

On behalf of ...... **C.U.STOMER** ......
(Insert applicants name if signed by Agent)

Don't forget to sign the forms.

On a full Planning form there is a Part 2 for non-domestic applications. Put two lines through the whole section and put N/A.

When submitting plans and forms the following numbers are usually sent in (look at your forms – some authorities have different requirements):

Planning | Building Control
Copies of Form 4 | Copies of Form 2
Copies of Plan 4 | Copies of Plan 2

With a Planning application a certificate must also be submitted. The most usual certificate are Certificates A and B.

Certificate A is used where no one other than the applicant has an interest in the land.

When you sign this form you are certifying the following:

1. That the applicant owned the land;

2. That none of the land was a farm.

Alternative

1. That you have notified tenants on the farm.

Certificate B is used when another party has an interest in the land.

When you sign this form you are certifying the following:

1. That you have notified all other owners of the land;

2. That none of the land was a farm.

Alternative

2. That you have notified tenants on farm.

N.B. This certificate should be issued when a wall is built on a boundary. If ordinary footings are used the foundations will project into the neighbour's garden. **This is legal if the neighbour agrees.**

However – you cannot build on a neighbour's land without permission.

A tip to save planning time

If you want to build on the boundary and you are using traditional foundations that project into the adjoining property then get the neighbour to give you a letter saying that he has no objection to your proposals.

You will still have to issue a Certificate B but if you send a copy of the letter with your Planning Application the Council will know in advance that there will be no 'neighbour' problems.

N.B. If you issue a Certificate B then a copy of Notice No. 1 must be sent to the owner of the neighbouring property.

If the neighbour will not let you undermine his land then you have to set your wall back minimum 150mm from the boundary or use a **raft.**

Fees

A set of Fee charges can be obtained from Planning and the Building Control Department.

At the moment for domestic extensions the Planning Charge is £33.00.

The Building Control charge has to be calculated from a schedule and depends either on floor area or costs.

A tip on fees

Always get your client to pay the Local Authority Fees. It saves your pocket and interest charges.

I normally ask for crossed cheques made out to the Local Authority. If the Local Authority don't agree with the fee they will notify you accordingly.

Typical conditions of engagement

As previously indicated, at the back of the manual, there are typical conditions of engagement. You may believe that these are not necessary or that you could create better terms of engagement yourself. I have placed them there to make you think. Conditions are never needed until you meet the bad client. That is when you need them. Study them carefully . . . I have built them up over the years.

Conclusion

I hope that I have been able to provide the knowledge, or direct you towards further reading matter, which will enable you to start your own Building Consultancy, or if you merely wish to prepare plans for yourself, help you avoid the common pitfalls.

There is one thing that this book will not provide and that is the will to win. You must provide this.

appendices

Appendix A

Most Councils issue guidelines for extensions to domestic properties in their area. I have indicated below a typical 'Design Policy' but this must not be taken as gospel. The policy may differ from Council to Council.

Note well

There is a basic anomaly in Planning in England and Wales which is that if your extension constitutes 'Permitted Development' then the Council has virtually no control over the design. In other words, generally speaking, Design Policies can only be enforced if you **need** planning approval.

It is quite possible to get a situation where a planning officer at a Council will not be able to approve your plans because it does not comply with the Council's design policies if you apply for full planning permission but if your extension complies with all the parameters of Permitted Development then you can still build it. My advice is to check to see if you need planning permission first.

Typical design policy document

Domestic extensions generally

1. Harmonising – all materials should match as closely as possible those of the existing structure.
2. Overlooking Neighbours – no proposed windows should be close to the neighbour's boundary.
3. Minimum distance to rear boundary – a rear extension must leave a minimum of 7.50 metres between the extension and rear boundary.
4. After construction of new extension – the rear garden should have remaining at least 50m² of free space.

Single storey extensions

5. Rear extensions adjacent to or directly on the party boundary – should not exceed 3.0 metres. An existing extension adjacent **may** allow an increase in depth to be considered.
6. Side extensions – these are acceptable **provided** that it does not preclude the provision of a garage/garage space behind the front most wall of the building.
7. As a general rule – any extension should not project beyond existing building lines.

Two storey extensions

8. Terracing effect – side extensions will not be allowed if they create a terracing effect between two properties.

9. Flat roofs are not acceptable on two storey extensions unless the existing property has a flat roof.

10. On semi-detached property/terraced dwelling – two storey extensions adjacent to a boundary will in most cases not be considered acceptable.

Dormer extensions

11. Where a loft is converted a dormer will not be acceptable if the new construction projects above the ridge line of the existing property.

Garages

12. Projection in front of the foremost part of the house – this will not usually be considered acceptable except where built in conjunction with front porch extension.

13. Where a garage is constructed it must have minimum dimensions externally of 5 metres and a width of 2.40 metres and there must be at least 5 metres left as driveway between the garage doors and the back of pavement line.

14. Unsightly bonding on front elevations – in order to avoid cutting bonding being visible on front elevations the garage must be set back at least 112mm where it is attached to the dwelling unless constructed in conjunction with a porch.

Garage Conversions

15. Where it is proposed to convert an existing garage into habitable accommodation this will only be considered acceptable where a garage space of at least 5 metres × 2.40 metres is provided behind the foremost wall of the original dwelling.

Porches

16. In most cases it will not be considered acceptable if the porch projects more than 1.50 metres from the foremost wall of the original dwelling.

Contract			Job ref.
Part of structure	**Raft Foundation**		Calc. sheet No. 1
Drawing ref.	Calculations by	Checked by	Date

Members ref.	CALCULATIONS	OUTPUT

For Loadings and References used,
see 'Loadings Page' over.

For Loadings and References used, see "Loadings Page" over

Raft Foundation.
Loading onto raft edge from roof　　　　 = 1.5 x 3/2　 = 2.25 kN/m
Load from blockwork & plaster (inner leaf) = (1.8 + 0.3) x 3 = 6.3　kN/m
　　　　　　　　　　　　　　　　　　　　　　　　　　　　　　　8.55 kN/m

Loading on outer bwk leaf = 2.4 x 3 = 7.2 kN/m
Consider loading over a 400 wide strip at edge.

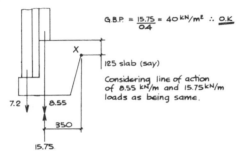

G.B.P. = $\frac{15.75}{0.4}$ = 40 kN/m² ∴ O.K

125 slab (say)

Considering line of action
of 8.55 kN/m and 15.75 kN/m
loads as being same.

7.2　　8.55

350

15.75.

Moment at X = (7.2 x 0.5) - (7.2 x 0.35) = 1.08 kN/m
Ultimate moment = 1.6 x 1.08 = 1.73 kN/m
M/bd^2 = 1.73 $10^6 / 10^3$ x 95² = 0.2
∴ As = 0.15% x 10³ x 95 = 142.5 mm²/m

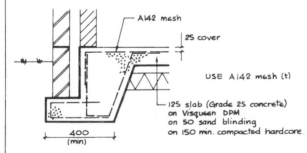

A142 mesh

25 cover

USE A142 mesh (t)

125 slab (Grade 25 concrete)
on Visqueen DPM
on 50 sand blinding
on 150 min. compacted hardcore.

400
(min)

Note : Cover to reinforcement to be 40 mm,
except where noted otherwise.

	Contract			Job ref.
	Part of structure	**Loadings**		Calc. sheet No. 2
	Drawing ref.	Calculations by	Checked by	Date

Members ref.	CALCULATIONS	OUTPUT

References

British Standard CP3 : Chapter V : Part 1 : 1967.
British Standard BS 648 : 1964.

Pitched Roof Loading

		kg/sq m
Interlocking tiles		50
Battens and counterbattens		7
Sarking felt		3
Roof rafters	say	13
Fibreglass insulation		2
Plasterboard and skim		20
Superimposed loading		75
		170

Design loading taken as 1.7 kN/m^2

Dormer Roof Loading

Chippings		20
Three layers of roofing felt		6
20 mm chipboard		15
Roof joists and firrings	say	10
Fibreglass insulation		2
Plasterboard and skim		20
Superimposed loading		75
		148

Design loading taken as 1.5 kN/m^2

Floor Loading

20 mm flooring grade chipboard		15
Floor joists	say	15
Superimposed loading		150
		180

Design loading taken as 1.8 kN/m^2

Partitions

Studding and fibreglass insulation	say	7
Plasterboard and skim		20
		27

Design loading taken as 0.3 kN/m^2

Appendix C

Appendix C

Notes regarding planning and building control procedure and conditions of engagement

(A) GENERALLY

In these conditions the words 'I' or 'We' or 'Us' means the surveyor or the practice of J Soap & Co.

The term 'you' or 'your' refers to the owner of the property or the person who has authorised the plans to be prepared.

The plans are purely for the use of you or your builder and is not for issued to third parties without permission and we will not be responsible for any alleged losses incurred should third parties act upon the details provided.

Normally (unless your house is listed or in a special area) the only approvals required when you build an extension are:

1. Planning Approval (or Class 1 Permitted Development);

2. Building Control approval.

Should Listed Building Approval etc be required this is an extra service. If your property is a listed building or in a special area or if there are tree preservation orders in force, it is essential that you let the Surveyor know during his visit so that he can make the necessary applications on your behalf. The responsibility for letting us know whether or not the property is listed or in a special area is your responsibility and no claims will be accepted at a later date if this information is not forthcoming. If you are in any doubt you must telephone your council and let us know as soon as possible.

DIMENSIONS ON THE DRAWING

Whilst every effort is made to ensure that the dimensions on the plan are accurate, errors can occur. As we are not a party to the contract between yourself and your future builder we make it a condition on our plans that your builder accurately checks all dimensions on site before starting work and before ordering materials.

This is specifically done so that we and you are protected against the possibility of an error getting through the checking procedure (see later for details).

We would stress that it is essential for you to insist that your builder carries out these checks before carrying out any work. Should the contractor need to make amendments you should tell the building control officer before carrying them out. We do not accept any claims for negligence or consequential loss with regard to incorrect dimensions shown on the plan as these could have been corrected, under the checking procedure, and the builder has been specifically instructed on the plans to check dimensions before proceeding.

FOUNDATION DETAILS

Dimensions given on foundations are only indicative for normal soil conditions. As trial holes have not been dug on site it is impossible for us to do otherwise unless advised by you that you know of bad ground in the area. If it is necessary to revise the foundations on site once the works commence this is not a defect on the plans and the plans specifically advise the builder to agree exact foundation details with the building control inspector. Should it be necessary to provide calculations or amend the drawings in the light of on site excavation this is an additional service and a charge will be made.

DRAINAGE DETAILS

All drainage shown on the plans is provisional and may require on site agreement with the building control officer.

(B) PLANNING APPROVAL/PERMITTED DEVELOPMENT

Normally planning permission is required for extensions to domestic property.

However in certain specific circumstances formal planning permission is not required. When we are of the opinion that the development constitutes permitted development, our normal procedure is to forward a copy of the plan to the Local Authority and request a formal confirmation that the Proposals are Class 1 permitted development.

Should the Local Authority require a formal application we would notify you accordingly.

The advantages of Class 1 approval are:

1. There is no planning fee.

2. It normally takes less time.

At the present moment the normal planning fee is £33.00 for domestic extensions but fees increase year by year with inflation. Should the fee increase while your application is being submitted we will notify you accordingly.

(C) BUILDING CONTROL

Most house extensions require Building Control approval. N.B. The plans when prepared must not be acted upon until they have been approved in accordance with clause 13 and 11(1)(b) of the Building Regulations 1985. Should the owner or builder commence work without the approvals they do so at their own risk.

The fee charges are on a sliding scale. At present the fee charges vary according to the size of the extension. The normal fee for small extensions is either £12.65 or £25.30 but the exact details for your extension will have to be calculated from the tables.

(D) DETAILS REQUIRED ON A PLAN

In order to be acceptable to the Local Authority your plans must have the following details:

1. Plans of Proposed and Existing at scales of min 1:100.

2. Elevations of Proposed and Existing at scales of min 1:100.

3. Sections through the property and at scales min 1:100.

4. Location plan   These can sometimes be combined.

5. Block plan   These can sometimes be combined.

6. Specification.

(E) TIME

Time is not the essence of the contract. Whilst approximate times are given below, once the documents are lodged with the council we are powerless to speed up the approvals although we do try to contact the individual surveyors at regular intervals in order to assist progress.

(F) APPLICATION TIME/MINIMUM TIME TO APPROVAL

Planning   2 Months

Building Control   5 Weeks
(Both times exclude the length of time taken to prepare the plans.)

THESE APPLICATIONS ARE NORMALLY SUBMITTED AT THE SAME TIME.

(G) SERVICE

The normal services carried out on your behalf are the initial survey of the property, preparation of working drawings at scales 1:100 and 1:50 with sufficient detail and information for interpretation of the proposed works for submission to the Local Authorities and for issuing to building contractors to obtain tenders or quotations.

(H) BOUNDARIES

We prepare the plans based upon dimensions taken on site. Where there are no obvious boundaries or they are hidden from view by debris, snow or existing buildings, we will agree the dimensions with you. Where fences and walls exist between properties we accept these 'natural boundaries' as being correct.

If you have any doubt regarding the ownership of any land we would advise you to speak to your neighbour and obtain a letter giving their consent to the proposals or contact your solicitor or building society for clarification.

(I) FEES NOT INCLUSIVE

Our fees are exclusive of all Local Authority charges (i.e. Planning and Building Control Charges), the cost of Consultant Engineers charges for preparation of structural calculations which may be required by Building Control.

(J) FEE CHARGES

An estimate will be provided, if required, for the likely cost of fee charges. The estimate is based upon normal hourly charges for surveyors and draughtsmans time. However, should your requirements or Local Authority create excessive demands on time, we reserve the right to revise charges as necessary. An invoice to cover fees will be issued to you immediately the plans are completed and your remittance should be forwarded with the Local Authority fees unless indicated otherwise on the confirmation of instructions. The Practice reserves the right to charge interest on unsettled invoices at a rate of 2% above the current National Westminster base rate.

(K) ARBITRATION

With the exception of you not paying your fee charge (see J above), if during the continuance of this contract or at any time thereafter any dispute, difference or question shall arise between the client and the surveyor in regard to the contract or the construction of these conditions or anything therein contained or the rights or liabilities of the client or the surveyor, such dispute, difference or question shall be referred pursuant to the Arbitration Act 1979 or later amendments to a sole Arbitrator to be agreed upon by the client and the surveyor and failing agreement to be appointed at the request of either party by the President for the time being of the Chartered Institute of Building or the Institute of Architects and Surveyors. The Arbitrator's decision shall be final and binding.

(L) ON-SITE SUPERVISION

Our normal charge does not include on-site supervision unless agreed otherwise. You are advised to check the builders work carefully as it proceeds . . . If in doubt about any section of work . . . Ask your builder . . . Make sure you get what you really want.

(M) PROVISION OF SAMPLE PLAN FOR APPROVAL

A sample plan will be provided for your approval approximately 14 days after the surveyor visits your house. Whilst every effort is made to ensure the accuracy of plans it is essential that you examine the plan carefully and ensure that the details are to your satisfaction. If you do not check your plan you could cause yourself a great deal of trouble on site if an error escapes detection. In particular you should check that the major dimensions are as you instructed. No claim will be accepted at a later date for alleged defect or negligence on our part and it is a specific condition of contract that at no time will the surveyors liability exceed the value of the fee charged. Once we receive your cheque for the Local Authority fees (see later) this will be taken as approval of the plans (unless you contact us and let us know what revisions are required). **There is no charge for revising the plans** as long as the changes required are reasonable requests (e.g. there would be no charge for altering a window dimension or door location etc).

(N) DISBURSEMENTS

The cost of printing drawings for your personal use will be charged at a current local printer's rate.

(O) V.A.T.

The practice is V.A.T. registered and statutory V.A.T. will be added to our fees.

(P) COPYRIGHT

In accordance with the provisions of the Copyright Act 1956 or later amendments, copyright in all drawings remain the property of J Soap & Co unless otherwise stated.

Appendix D

'ON SITE' FORM FOR CLIENTS SIGNATURE

I confirm that the Conditions of Engagement have been explained to me and that I have been handed a copy for my retention.

I accept the estimated figure of £   for the services required and understand that this figure will not necessarily be the final cost in the matter.

I enclose my deposit of £   and in so doing formally request that you commence the services as outlined.

I understand that I will receive a formal confirmation of instructions and a copy of the plans prior to submission to the Local Authority. Should there be any minor amendments that I require to the plan I understand that these will be made at my request at no extra charge as long as these amendments are made prior to submission.

We understand that before the formal submission can be made we must forward the Local Authority fees as requested in the Confirmation of Instructions.

SIGNED ....

NAME ....

ADDRESS ....

....

POST CODE .... DATE ....

TELEPHONE ....

STANDARD LETTER

REF/JS/SL2/

Dear Mr and Mrs Jones,

Proposed work at your house

I have pleasure in enclosing a copy of your plan as promised, a copy of my confirmation of instructions and a further copy of our Conditions of Engagement.

Should the plan require any amendments I would appreciate it if you would contact me as soon as possible so that valuable submission time is not lost.

It is essential that you check your plan over carefully, in particular the major dimensions, and ensure that everything is as instructed.

I have prepared your plans based upon dimensions taken on site to 'Natural Boundaries' (e.g. Fence Lines). If you are in any doubt regarding dimensions indicated on our plan then I would be obliged if you could contact me as soon as possible.

As agreed, my fee does not include for on-site supervision of the works.

Would you please forward the Local Authority fees as soon as possible, and I will then be able to submit the application on your behalf. I will be forwarding my fee account to you for your attention under separate cover.

Yours sincerely,

J. Soap

REF JS/SL3/

Confirmation of Instructions

Client: Arthur Daily

Address: Somewhere, London.

Service: **Prepare plans to client's approval and submit to Local Authority. No on-site supervision allowed.**

Agreed **fee** £   (excludes planning and building regulation fee charges; any engineering fee charges which are or may be required by the Building Inspectors; on-site supervision of the works, and V.A.T. of £   ).

Please forward a cheque for £12.65/£25.30 for building control.

Please forward a cheque for £33.00 for planning.

Note: My fee account will be forwarded to you under separate cover.

The Building Control and Planning charges are to be sent to this office prior to submission of the application. Should the Building Control fee be considered too low or too high, you will be notified by the Local Authority direct, and the details of credit/extra charge will be supplied.

The Chief Technical Officer

REF JS/SL4/

Dear Sirs,

Project:

We enclose herewith plan JS/ in connection with the above.

We are of the opinion that the work detailed on this plan constitutes class 1 permitted development.

We would be obliged if you would confirm in writing that planning permission will not be required in this case. We have however formally applied to the Building control department for your approval.

Yours faithfully,

J. Soap.

REF JS/SL5/

Dear Mr

Proposed works at your house

We have pleasure in enclosing your planning approval. We will forward Building Control approvals as soon as we receive them.

It is essential that you check your plan over carefully. In particular the major dimensions, and ensure that everything is as instructed before giving the plans to your builder.

Note well:

We have prepared our plans based upon dimensions taken on site to 'Natural Boundaries' (e.g. Fence Lines and Walls). If you are in doubt regarding your boundaries you must contact your solicitor or building society for clarification before proceeding.

If you are in any doubt regarding any other or dimensions or details indicated on our plan it is essential that you contact us before the work is put in hand.

We trust, however, that we have interpreted your instructions correctly and that you will have no problems in the construction stage.

Yours sincerely,

J. Soap.

REF SL9/ BUILDING CONTROL 11(1)(b) SUBMISSION

Dear Sirs,

Proposed works at 'Underwood', Somewhere.

Please find enclosed herewith documents in connection with a **'Full Plan'** application under regulations 11(1)(b).

Should there be any queries would you please telephone my office as soon as possible in order that delays can be avoided.

Yours faithfully,

J. Soap.

Appendix F

SUGGESTED SPECIFICATION FOR SMALL EXTENSIONS

Standard specification of works (August 1987)

This plan the copyright of J. Soap and Co., and is not to be reproduced without permission and is for use as a planning and building control document. All building construction is to comply with the Building Regulations 1985 (or later amendment) and the approved documents A to L, and mandatory rules for means of fire escape. These plans shall not be acted upon until they have been approved in accordance with clause 13 and 11(1)(b) of the Building Regulations 1985. Should the owner or builder commence work without the above approval they do so at their own risk.

The term Builder or Contractor shall mean the person responsible for the construction of the works. The contractor shall ensure that a responsible person is on site during normal working hours to take instructions. The Contractor is advised to visit the site, prior to quoting and to make due allowance when preparing his estimate for access, availability of labour, plant and all things necessary for the construction of the works. No claim will be accepted for want of knowledge at a later date.

Give all Notices to Local Authority and Public Undertakings and pay all fees and charges. Likewise the Contractor shall include for all costs arising from compliance with all Statutory Orders, Regulations, Building Regulations, Bye-Laws and any Acts of Parliament.

The Contractor shall protect the premises during the execution of the work against all damage or vandalism and shall provide tarpaulins and all other necessary coverings, and take adequate precautions to keep new and existing work free from damage by inclement weather during the progress and clear away on completion. The premises are to be secure at the end of each day's work, and the Contractor must reinstate at his own expense any damage caused by neglect in protecting the building.

The drawing and specification is to be read as a whole; if any details whatsoever are not clearly shown or specified the Contractor is to ask for instructions, and if any work be wrongly done, it shall, if the Surveyor so directs, be removed and done again at the Contractor's expense.

Site copies of the drawings must be available on the site during the progress of the works for inspection by the building inspector/owner/surveyor.

All dimensions given whether figured or scaled are to be physically checked on site by the contractor prior to commencement of work and the contractor will take responsibility for same. **Any anomalies are to be reported to the surveyor prior to the work being put in hand and prior to ordering materials. If in doubt . . . ask.** Figured dimensions are to take preference over scaled dimensions but scaled dimensions are not to be ignored. The Contractor must furnish the Local Authority with notices of commencement of work and stage of completion and must liaise with water, gas, electricity and British Telecom as necessary and comply with their requirements.

The Contractor will be required to maintain and protect all gas and water pipes, electricity cables, sewers, etc, and other public property or property of the Local Authority or Public Utility Company which may be encountered during the progress of the works and he shall be responsible for and properly make good any damage to the same to the satisfaction of the Authorities concerned.

Deviations from the drawing can be made but only with consent to the client, building inspector and surveyor.

If during the continuance of the Contract or at any time thereafter any dispute, difference or question shall arise between the client and contractor in regard to the contract or the construction or the rights or liabilities of the client or the contractor, such dispute, difference or question shall be referred persuant to the Arbitration Act 1979, to a sole Arbitrator to be agreed upon by the client and the contractor and failing agreement to be appointed at the request of either party by the President for the time being of the Chartered Institute of Building. The Arbitrator's decision shall be binding and final.

No part of the work shall be sub-let to other persons unless the written authority of the **owner and surveyor** is obtained.

Dimensions given on foundations are only indicative for normal soil conditions. Should it be necessary to provide raft foundation or other construction the builder shall contact the surveyor as soon as possible. All goods and materials unless otherwise specified, shall be in accordance with the latest British Standard and Code of Practice current at the date of tendering.

All materials, appliances, fittings, etc, must be obtained from sources approved by the **owner and surveyor** with reasonable samples of materials to be used in the work, which samples if approved, shall become a standard of quality.

All building construction is to comply with the requirements of the Registered Housebuilder's Handbook as published by the National Housebuilding Council and the **requirements of the Public Health Acts** and other acts of Parliament. **In particular, the contractor shall pay regard to Local Authority requirements in connection with clearing bins and access to rear gardens.**

The Contractor is to carefully check the boundaries of the site and is not to build on land not owned by the client without obtaining the neighbour's consent.

The Contractor must indemnify and insure the owner for any damage to persons and/or property for the sum of not less than £500,000 and the Contractor will be held liable for any damage or nuisance to, or trespass on the adjoining property arising from or by reason of the execution of the work, and he must take all necessary steps to prevent any such trespass or nuisance being committed.

No trial holes have been taken on site and the builder must acquaint himself with the ground conditions of both the site and adjoining areas. Comply with NHBC Practice note No. 3 'Root Damage by trees – siting of dwellings and special precautions'.

All brickwork shall comply with the recommendations contain in NHBC pocket book issue No. 1.

Provide adequate DPC's to new brickwork. DPC's to be a minimum of 150mm above adjoining ground level. Returns to windows and door openings in new cavity brickwork to have vertical DPC.

All windows to habitable rooms shall be one-tenth of the floor area and shall have opening vents size not less than one-twentieth of the floor area.

Roof joints to be strapped using 30mm × 5mm galvanised mild steel straps at 1.8m centres in accordance with the building regulations.

All elements of structure shall have minimum half-hour fire resistance. This will be achieved using two layers of 9.5mm plasterboard and skim.

Foundation and floor slab alternative 1

Concrete strip foundation 1:3:6 mix (sulphate resistant if necessary) and aggregate to comply with BS 882 to be taken down a minimum depth of 900mm from ground level. The concrete foundations shall be 600mm wide × 150mm thick minimum size. Should the Building Inspector instruct that differing sizes are to be constructed, the contractor is to comply with such instructions. All foundations to be taken down to good bearing strata and in accordance with Local Authority requirements.

The concrete floor slab or oversite concrete under timber floors shall be minimum 100mm and be 1:2:4 mix (sulphate resisting if necessary).

Foundation alternative 2

The raft foundation shall be minimum 150mm thick and shall be 1:2:4 mix (sulphate resisting if necessary).

Floor finishes

To solid floors in habitable areas provide 50mm cement and sand screed (1:3) or alternatively 20mm polished asphalt to receive owner's floor finish. Asphalt for flooring shall comply with the recommendations of the Mastic Asphalt Council's booklet 'Application of Mastic Asphalt'.

The quality shall be Flooring Grade. Provide and install Visqueen 1200 gauge DPM to underside of floor slabs oversite concrete or rafts (if applicable) laid on 100mm bed of hardcore blinded with sand. The Visqueen shall be turned up and dressed under DPC in walls at abutments. The floor slabs oversite concrete or rafts (if applicable) shall not be laid lower than the adjacent ground level. If existing levels are inconsistent with this contractor must lower adjacent levels accordingly.

External walls

The external walls to new habitable areas shall have a minimum 'U' value of 0.6 and comprise facing bricks (or thermal blocks and render if applicable) to match existing, 50mm cavity with vertical twist ties complying with BS 1243:1972 spaced at 900mm centres horizontally (max) and 450mm vertically and additional ties at openings and filled with 'Dritherm' fixed in accordance with NHBC recommendations and an inner leaf of 100mm approved thermal blocks. Alternatively the contractor can replace the 50mm 'Dritherm' and 100 blocks with 50mm cavity (unfilled) and Thermalite 125mm Turbo blocks **providing that suitable 125mm blocks are also used as support in substructure.**

Where cutting and toothing is not possible 'fir fix' stainless steel connections shall be used.

The contractor must ensure that ends of beams are adequately seated on suitable padstones and that lightweight blockwork is not over stressed. The heads of all cavity walls are to be closed. Returns around door and window openings shall be closed with brickwork or blockwork and incorporated a DPC. Where exposed to view all returns in facing brickwork shall be in facings.

Flat roofs

The flat roof shall be of the cold type and contain 100mm glass fibre and shall comprise of 12.7mm white limestone chippings on 3 No. layers of Bituminous felt. **The roof deck** shall either be WBP 18mm plywood or chipboard (roofing grade). **The contractor is to inform the client of the extra over cost of using plywood.**

On no account must standard chipboard be used. The contractor will be expected to ensure that the chipboard (if used) is to BS 5669:1979 and type 3 or type 2/3 and marked with BS colour stripes (green denoting type 3 and red/green type 2/3).

Falls are to be created using softwood firrings laid to fall not less than 1 in 40 Softwood joists are to be at maximum 450mm centre. Provide 50mm × 50mm softwood splayed tilting fillets at edges of the roof in accordance with good building practice. On all types of chipboard (whether prefelted or not) the contractor shall provide three layers of felt. The bottom layer shall be partially bonded and be to BS 747 type 3G or 2B. The second and third layers shall be either type 2B or 3B in accordance with NHBC recommendations.

Pitched roofs

The roof is to be covered with Marley Modern tiles (or other equal and approved interlocking roof tiles) laid in strict accordance with the manufacturer's instructions given in 'Technical Roofing Manual', including recommended roof fixings and pitches. Mortar used on the ridge and verge shall match the colour of the tile. The following are specific details.

1. The tiles are to match the colour of the main roof.

2. The roof pitch shall not be less than 23 degrees for smooth tiles and 25 degrees for granular tiles (unless otherwise detailed).

The reinforced underslating 'Andersons Twinplex' type 1F or other equal and approved, is to be fixed to the s.w. rafters with 2 No. galvanised nails size 20mm into each rafter. The felt is to have 150mm headlaps, and is to lap 50mm into the gutter.

Insulation, ventilation etc

Provide 100mm fibreglass insulation to roof over habitable areas.

Unless indicated to the contrary on the drawing, ventilation of roof voids to be provided using Glidevale ventilation as manufactured by Willian Building Services (Tel: 061 973 1234). Roof ventilation is to comply with Building Regulations and BS 5250. Where cross ventilation is not possible in voids in flat roofs in enclosed corner situations the contractor shall install 'Anderson' or similar 'Mushroom' type ventilators on the roof surface to provide ventilation.

The contractor shall install all necessary soakers, flashings, aprons, and the like at all abutments sufficient to prevent water entering the building, and for plasterwork, skirtings, architraves, windowboards, etc, **the contractor shall install all electrical and plumbing work and fittings (if applicable) in strict accordance with the best building practice exact details of electrical and plumbing (where not detailed) shall be agreed with the client prior to commencement.** Electrical installations shall comply with the IEE wiring regulations Fifteenth Edition 1981 or later edition if applicable.

New manholes (where provided) will comprise 150mm concrete bed with benching to channels, 255mm class B engineering bricks in cement mortar 1:3 150mm reinforced concrete cover slab and mild steel cover. Drains shall be hepseal or hepsleeve flexible jointed 100mm clay pipes with 150 beds and surround laid in accordance with hepworth recommendations and to falls to comply with the Building Regulations.

Encase all drains under extension with 150mm concrete

Internal dimensions for inspection chambers

UP TO 1200mm INVERT DEPTH		1200-1800mm INVERT DEPTH	
100mm Main Channel Length	150mm Main Channel Length	100m Main Channel Length	150mm Main Channel Length
1Br 600mm	1Br 600mm	1Br 800mm	1Br 800mm
1Br 600mm	1 2Br 800mm	2Br 910mm	2Br 1100m
3Br 910mm	3Br 1150mm	3Br 1300mm	3Br 1600mm
Width		Width	
Branch from one side only		Branch from one side only	
450mm	700mm	800mm	800mm
Branches from both sides		Branches from both sides	
700mm	700mm	1000mm	1100mm

Allow for keeping the works clear of rubbish during the currency of the Contract and remove from time to time all debris as it accumulates and leave clear and tidy on completion to the satisfaction of the owner.

Appendix 'G'

Typical advisory note likely to be issued by your local council

Questions and answers on new 1985 building regulations

Question One

I know that there are some differences between the old regulations and the new – is it correct that I no longer need to submit plans for small extensions?

Answer

The law is that you must submit a building notice if you want to carry out alterations and extensions to your property. This can be done in one of two ways.

1. Using the existing method and depositing a full set of plans.

2. Use the 'No – plan option' but you must notify the council of your intentions and submit a block plan and specified information.

Question Two

Do I need to get in touch with the Council before I start work?

Answer

It is an offence to carry out work without notifying the Council. You are also required to give 48 hours notice prior to commencing work so that the Building Inspector can make arrangements to visit your house.

Question Three

As the Building Regulations have changed, have all the other regulations changed too or do I still need planning permission and the like.

Answer

The new Building Regulations have not altered any other Regulation (e.g. Tree Preservation Orders, Listed Buildings, etc).

Question Four

Will the no plan option cost less?

Answer

No – the total fees chargeable by the Council are the same.

Question Five

Which option is best for me?

Answer

Sometimes full plans have to be submitted anyway. Using the 'No-plan' option can save time but if you contravene the Building Regulations then the work might have to be pulled down. The traditional full plan option is safer.

Question Six

Do I need permission to get my cavity walls filled with foam or other insulation?

Answer

Yes.

recommended books

Recommended list of books/leaflets for further reading

Name	*Publisher*
THE BUILDING REGULATIONS (November 1985) and approved documents	H.M.S.O.
THE BUILDING REGULATION EXPLAINED AND ILLUSTRATED – SEVENTH EDITION VINCENT POWELL-SMITH and M. J. BILLINGTON	COLLINS
HOME EXTENSIONS – FLAT ROOFS	FIBREGLASS ST. HELENS
REGISTERED HOUSE BUILDERS HANDBOOK 1974 (with amendments 1983/84) including Practice Notes 5, 6, 8, 11, 12 and 13	THE NATIONAL BUILDING COUNCIL 58 Portland Place London W1N 4BU
BUILDERS REFERENCE BOOK	BTJ BOOKS
BUILDING REGULATIONS IN DETAIL JOHN STEPHENSON	BTJ BOOKS